THE THEORY OF ALMOST EVERYTHING

ROBERT OERTER teaches physics at George Mason University. He received his Ph.D. from the University of Maryland. He has done research in the area of supergravity, especially as applied to super-string theories, and in the quantum mechanics of chaotic systems. He lives in Maryland.

"In an era when enormous attention is being paid to the promising but highly speculative superstring/M-theory, a great triumph of science has gone nearly unnoticed, except by physicists. Robert Oerter provides here an accessible introduction to the Standard Model—a towering example of human creativity. He outlines how the Standard Model can serve as the launching pad for humanity to—paraphrasing Einstein—see better the secrets of 'the Ancient One.' "
—S. J. Gates Jr., John S. Toll Professor of Physics and director of the Center for String and Particle Theory, University of Maryland

"We always hear about black holes, the big bang, and the search for life in the universe. But rare is the book that celebrates the Standard Model of Elementary Particles—a triumph of twentieth-century science that underpins nearly all we know about physical reality. Oerter's *The Theory of Almost Everything* belongs on anyone's shelf who cares about how the universe really works."
—Neil deGrasse Tyson, astrophysicist and author of *Origins: Fourteen Billion Years of Cosmic Evolution*

"The tried and true Standard Model of particle physics isn't getting the respect it deserves. Physicist Robert Oerter is out to set the record straight, and in this book he reveals the Standard Model in all its glory." —*New Scientist*

"Accessible and engaging . . . This book is for anyone interested in modern physics and ultimate answers about the universe." —*Science News*

"The material is deep and rich. The exposition is very clear . . . interested laypersons will find this book rewarding reading." —*Choice*

"This highly accessible volume explains the Standard Model to the everyman, using literary references and easy-to-follow analogies to make clear mind-bending physics principles." —*Publishers Weekly*

The Theory of Almost Everything

The Standard Model,
the Unsung Triumph of Modern Physics

Robert Oerter

A PLUME BOOK

PLUME
Published by Penguin Group
Penguin Group (USA) Inc., 375 Hudson Street, New York, New York 10014, U.S.A.
Penguin Group (Canada), 90 Eglinton Avenue East, Suite 700,
Toronto, Ontario, M4P 2Y3, Canada (a division of Pearson Penguin Canada Inc.)
Penguin Books Ltd., 80 Strand, London WC2R 0RL, England
Penguin Ireland, 25 St. Stephen's Green, Dublin 2, Ireland
(a division of Penguin Books Ltd.)
Penguin Group (Australia), 250 Camberwell Road, Camberwell,
Victoria 3124, Australia (a division of Pearson Australia Group Pty. Ltd.)
Penguin Books India Pvt. Ltd., 11 Community Centre, Panchsheel Park,
New Delhi – 110 017, India
Penguin Books (NZ), 67 Apollo Drive, Rosedale, North Shore 0632,
New Zealand (a division of Pearson New Zealand Ltd.)
Penguin Books (South Africa) (Pty.) Ltd., 24 Sturdee Avenue,
Rosebank, Johannesburg 2196, South Africa

Penguin Books Ltd., Registered Offices: 80 Strand, London WC2R 0RL, England

Published by Plume, a member of Penguin Group (USA) Inc. Previously published
in a Pi Press edition.

First Plume Printing, October 2006
9 10 8

Copyright © Robert Oerter, 2006
All rights reserved

℗ REGISTERED TRADEMARK—MARCA REGISTRADA

CIP data is available.
ISBN 0-13-236678-9 (hc.)
ISBN 978-0-452-28786-0 (pbk.)

Printed in the United States of America

In Memory of George William Oerter

Contents

Introduction

> People are always asking for the latest developments in the uni-
> fication of this theory with that theory, and they don't give us
> a chance to tell them anything about one of the theories that
> we know pretty well... What I'd like to talk about is a part of
> physics that is known, rather than a part that is unknown.
> —Richard Feynman, *QED:*
> *The Strange Theory of Light and Matter*

There is a theory in physics that explains, at the deepest level, nearly all of the phenomena that rule our daily lives. It summarizes everything we know about the fundamental structure of matter and energy. It provides a detailed picture of the basic building blocks from which everything is made. It describes the reactions that power the sun and the interactions that cause fluorescent lights to glow. It explains the behavior of light, radio waves, and X rays. It has implications for our understanding of the very first moments of the universe's existence, and for how matter itself came into being. It surpasses in precision, in universality, in its range of applicability from the very small to the astronomically large, every scientific theory that has ever existed. This theory bears the unassuming name "The Standard Model of Elementary Particles," or the "Standard Model," for short. It deserves to be better known, and it deserves a better name. I call it "The Theory of Almost Everything."

The Standard Model has a surprisingly low profile for such a fundamental and successful theory. It has deeper implications for the nature of the universe than chaos theory, and unlike string theory, which is purely speculative in nature, it has a strong experimental basis—but it is not as widely known as either. In physics news items, the Standard

Model usually plays the whipping boy. Reports of successful experimental tests of the theory have an air of disappointment, and every hint of the theory's inadequacy is greeted with glee. It is the Rodney Dangerfield of physical theories, it "don't get no respect." But it is, perhaps, the pinnacle of human intellectual achievement to date.

Some of the Standard Model's architects are perhaps more visible than the theory itself: the clownish iconoclast Richard Feynman and the egotistical polymath Murray Gell-Mann have both written and been the subject of books. Many other names, though, are practically unknown outside specialist circles: Sin-Itiro Tomonaga, Julian Schwinger, George Zweig, Abdus Salam, Steven Weinberg, Yuval Ne'eman, Sheldon Glashow, Martin Veltman, Gerard t'Hooft. Perhaps part of the reason for the Standard Model's neglect is the sheer number of people involved. There is no solitary, rejected genius—no Einstein working alone in the patent office, no theory springing full-blown into existence overnight. Instead, the Standard Model was cobbled together by many brilliant minds over the course of nearly the whole of the twentieth century, sometimes driven forward by new experimental discoveries, sometimes by theoretical advances. It was a collaborative effort in the largest sense, spanning continents and decades.

The Standard Model is truly "a tapestry woven by many hands," as Sheldon Glashow put it.[1] It is, in this, a much better paradigm for how science is actually done than is the myth of the lone genius. But it conflicts with our prejudices about science and with the way popular physics is usually presented.

News reports and general-interest books about the Standard Model often emphasize the particles in the theory: the discovery of quarks, finding the W and Z bosons, looking for neutrino mass, the search for the Higgs particle. This emphasis misses the underlying structure of the theory. It is as if you were asked to describe a Christmas tree and you talked entirely about the ornaments and the lights, and never mentioned the tree itself: the piney smell, the color of the needles and the bark, the feathering of the branches, the symmetrical shape.

To a theoretical physicist, the quarks, electrons, and neutrinos are like the ornaments on the tree. Pretty, yes, but not what's fundamentally important. The structure of the theory itself is what's really fascinating. The Standard Model belongs to a class of theories called *relativistic quantum field theories.* Dream up any set of particles you like and you can write a relativistic quantum field theory to describe it. (I'll show you how to do this in Chapter 9.) All such theories incorporate the strangeness both of special relativity, with its paradoxes of time and motion, and of quantum mechanics, with its fields that are neither wave nor particle. The framework of relativistic quantum field theory adds some weirdness of its own: particles that pop into existence out of pure energy, and disappear again, literally in a flash of light. This structure encodes the rather bizarre worldview of the physicist. It tells what can be known about the universe, and what must remain forever mysterious. This structure, the deep symmetries of the universe that are hidden within this structure, and its implications for our understanding of the physical world, are what I want to tell you about in this book.

There is symmetry all around us: the shape of a snowflake or a daffodil, the crystal symmetry of a perfectly cut diamond, the volcanic beauty of Mount Kilimanjaro. We desire symmetry. Architects, artists, and composers incorporate symmetry into their creations. We judge faces with symmetric features more beautiful. When selecting a Christmas tree, the buyer walks all around it to see if it is attractive from all sides. And yet, too much symmetry is boring. A well-proportioned house is beautiful, but an endless row of identical houses is repellant. A musical phrase repeated over and over becomes monotonous, loses our interest, and soon becomes annoying. In a Jackson Pollock painting, one section of the canvas looks much like any other section, but no two areas are identical. Symmetry need not be perfect, it must not be perfect, to achieve beauty. As Francis Bacon said, "There is no excellent beauty that hath not some strangeness in the proportion".[2]

A tree, for example, displays many kinds of symmetry, not all of them obvious at first glance. Draw an imaginary vertical line through the center of the tree and you split it into two halves, each of which is a mirror image of the other, albeit an imperfect one. Symmetry of another kind can be found in the branching structure of the tree limbs. This structure is repeated for smaller branches, then for twigs, creating a kind of symmetry of scale. Select a small portion of a photograph of the tree and blow it up, then select a portion of the enlargement and blow that up. Each time the new photograph looks very similar to the previous one. This branching structure is repeated underground in the tree's roots, making the bottom half of the tree a distorted mirror image of the top half.

Symmetry can be destroyed. A building collapses in an earthquake; a wine glass shatters when dropped. A tree, buffeted by wind, falls over. When you walk around the fallen tree, it no longer looks the same from every side. Its crown is crushed by the ground: now, when you draw a line through the trunk, the two sides are no longer mirror images.

The story of fundamental physics in the twentieth century is a story of symmetry: symmetry perfect and imperfect, symmetry discovered and symmetry destroyed. The symmetries involved are not ones that can be seen with the naked eye, however. To discover them we must dive into the tree's inner structure. Its wood, viewed under a microscope, is made of cells, the cells built up of chains of molecules. The molecules in turn consist of atoms, which are constructed from still smaller particles. In a process of discovery that lasted the entire twentieth century, physicists learned that these smallest constituents of matter have symmetries of their own. If we could reach into the atom and give each of the particles a certain kind of twist, and if we could simultaneously give the same twist to every other particle in the universe, the world would go on exactly as if we had done nothing.

With a perfectly symmetrical face, you can't tell if you are looking at a photograph or a mirror image. The deep symmetries of the fundamental particles are exact—there is no way to tell if the twist has been made or not. Beyond these exact symmetries, not visible even in the fundamental particles but hidden in physicists' theories, lies yet another symmetry, one that existed in the first moments of the universe's existence, but has since been shattered. This symmetry and its downfall is the reason that matter as we know it exists, the reason for stars, planets, daffodils, and you and me.

The Standard Model is a theory of *almost* everything. Specifically, it is a theory of everything except gravity. Gravity may seem to be a major omission; in everyday life, gravity is certainly the force we *feel* most strongly. Without magnetism, the photos of your niece would fall off the refrigerator; without electricity, you could walk across a rug on a dry day and not get shocked when you touch the doorknob; but without gravity, you would go floating up off the Earth into space and asphyxiate.

Paradoxically, gravity is more noticeable to us because it is the *weakest* force. A proton, for example, has the smallest electric charge that it's possible to isolate in nature, yet the electric force between two protons is immensely larger (by a factor of 10^{36}!) than the gravitational force between them. Because the electric force is so strong, matter tends to hang out in neutral clumps, with equal amounts of positive and negative charges. The positive and negative charges cancel each other, and the resulting neutral clump doesn't feel any electric force from other neutral clumps. This is why we never see, for instance, an apple flying up out of a tree due to electrical repulsion from the Earth. The Earth is nearly neutral, apples are nearly neutral, so the net electric force is small compared to the gravitational force. Whenever an imbalance of charge is created, as when you shuffle across a rug, picking up extra negatively charged electrons from the rug, the imbalance will correct itself at the first opportunity. When you touch the doorknob, the extra electrons try to escape your body, repelled not by your personality, but by their mutual electric force and attracted toward any extra positive charges in the doorknob. The same thing happens in lightning strikes, when a large amount of charge flows back to the Earth from an electrically charged cloud, restoring electric neutrality.

Electric and magnetic forces, however, are much more important in everyday life than refrigerator magnets and static cling. The electric engine that runs the refrigerator contains magnets and uses electricity, as do the engines for your vacuum cleaner, your weed whacker, and your car's starter. Electricity flows whenever you turn on a light, a TV, a stereo, pick up the phone, cook on an electric range, or play electric guitar. Light is an electromagnetic effect, whether it comes from a light bulb or from the Sun. Your nerves send electric signals, so by the act of reading this sentence, you are causing a multitude of electrical events in your brain and your body. What's more, all chemical reactions may be traced to the electric and magnetic interactions of the atoms and molecules involved. Your body operates by way of chemical reactions, so electric forces are ultimately responsible for your movement, digestion, breathing, and thinking. It is electric forces that hold matter together, so the chair you are sitting in would not exist without electric forces. Far from being irrelevant for everyday life, electric and magnetic forces, together with gravity, *are* everyday life, or at least they are the substrate that makes life possible.

The Standard Model contains a complete theory of electric and magnetic forces, together with a description of the particles on which the forces act: protons, electrons, neutrons, and many more that are not as well known. So, in a sense, the Standard Model "explains" all those everyday phenomena, from the structure of the chair you sit on, to your very thoughts. It is not possible, though, to write an equation that describes your chair using the equations of the Standard Model (much less an equation for your thoughts!). The Standard Model equations can only be solved in very simple cases, say one electron interacting with one proton. In those simple cases, however, the Standard Model gives us such incredibly accurate predictions that we have a great deal of confidence that that is really how electrons and protons behave. (Other parts of the Standard Model, for instance the internal structure of the proton, are still not solved, and so our confidence is somewhat less for those areas.) Even though we cannot *in practice* use the Standard Model to describe a chair, we can say that a chair consists of protons, neutrons, and electrons in various configurations, and so, *in principle*, the Standard Model "explains" the chair at its most fundamental level.

Consider a computer as an analogy. The computer is made of wires, integrated circuits, a power supply, and so forth. Fundamentally, all that is "really" happening in a computer is that little bunches of electrons are being shuffled around through these circuits. However, when your computer tells you "ERROR 1175: ILLEGAL OPERATION, APPLICATION WILL BE SHUT DOWN," it is not very useful to pull out the circuit diagram for your CPU. Although it is possible *in principle* to describe what happened in terms of the circuits ("When memory locations A, B, and C have such-and-such a number of electrons, and some other number of electrons come down wire Q, then..."), this description would be useless for avoiding the problem. Instead, you need to be told something like, "Your operating system will only let you open four programs at a time. Shut down the excess programs before starting this one, and you won't get that error message." We can't locate "the operating system" or "program" on the circuit diagram—it is a higher level of description. Can we understand the error message by looking at the circuit diagram? No. Can we really understand the operation of the computer without understanding the circuits? No again. (Try building your own computer using only the Windows 2000 reference manual!) Both levels of description are necessary to "understand the computer," but the higher-level (operating system and program) functions can be explained in terms of the lower-level (circuitry) processes, and *not the other way around*. This is why we call the lower-level description the more fundamental one.

The Standard Model describes the "circuitry" of the universe. We can't understand everything in the universe using the Standard Model (even if we omit gravity), but we can't really understand anything at the most fundamental level without the Standard Model. Suppose you are a biologist who wants to understand the function of blood in the body. You need to investigate the penetration of oxygen across membranes, and its uptake by hemoglobin. Your biological question turns out to depend on chemical questions. To understand how fast oxygen is fixed by hemoglobin, you need to know about the configuration of the electrons in the oxygen and hemoglobin molecules. These configurations are determined by the electric and magnetic forces between the electrons and the nuclei, in other words, by the Standard Model.

To tell the story of the Standard Model and its symmetries, this book will follow a roughly chronological sequence. The reader should not be misled by this into thinking I am writing a history of the Standard Model. My goal is to give the reader an understanding of the theory itself. To give an accurate historical picture of the development of the theory, with all of the vagaries of blind theoretical alleys and inconclusive or incorrect experiments, would take us too far from our main goal. I have included some of the history so that the reader can understand the motivation for each new step taken, and to emphasize that the theory was developed in response to specific new discoveries about the way particles behave. It was not invented out of whole cloth by some theorist isolated in an office, but was painstakingly pieced together from the hints that experimenters managed to tease out of nature. The chronological approach may, at times, give a mistaken impression that the Standard Model developed by an orderly series of experimental and theoretical advances. This is far from the truth: the actual historical development was much more messy and interesting than I can convey here. The interested reader should consult the suggestions for further reading at the end of this book.

The story of the Standard Model must begin with the nineteenth-century worldview. Decades of careful experimentation had convinced physicists that everything that happened in the universe was a result of the interaction of particles and fields. Everything material, be it solid, liquid, or gas, consisted of unimaginably small particles, the atoms. They were pictured as tiny billiard balls moving in straight lines unless a force acted on them. A particle was endowed with the capability to generate a field filling all the space around it and influencing the motion of other particles. All forces arose from these fields. Particle generates field, field influences particle: this was all that ever happened in the universe.

The nineteenth-century worldview, known as *classical physics*, was stirred but not shaken by the discovery of a new symmetry in 1905. In everyday experience, space and time are completely different. We can move about in space; we can return home as many times as we like. Time, on the other hand, moves inexorably forward; there is no return. Albert Einstein's theory of special relativity forced physicists to change their perception of space and time. They are intricately intertwined—they are, in fact, two aspects of a single reality, which was termed *spacetime*.

Classical physics was shaken to its foundations by another set of discoveries around the turn of the century. The odd couple that triggered the earthquake was radioactivity and neon lights. According to quantum physics, which was developed to deal with the new phenomena, particles sometimes behaved like waves, as if they were not small and hard but spread out like a field. At the same time, fields could behave like particles. The two entities, particles and fields, that had seemed so different were starting to show a family resemblance.

By mid-century, physicists had successfully woven together the old, classical field idea and the new theories of special relativity and quantum mechanics. The framework that emerged from this union, known as relativistic quantum field theory, would prove remarkably robust. Indeed, it would be the framework used for fundamental physics for the rest of the century and the language in which the Standard Model would be expressed.

The discovery of quarks, hidden inside protons and neutrons, led to the discovery of a new and unsuspected symmetry, a new kind of "twist" of the quarks that leaves the world unchanged. This symmetry, called *color symmetry*, is intimately connected with the force that binds quarks into protons and neutrons, the strong force.

The great breakthrough that made a Theory of Almost Everything possible came with the realization that a symmetry (of the color symmetry type) could break spontaneously, just as a tree can spontaneously fall. We will learn how spontaneous symmetry breaking allowed physicists to predict the existence of new, never before observed particles. The discovery of all but one of those particles, with precisely the properties predicted by the theory, was the crowning achievement that confirmed the Standard Model. This downfall of symmetry is responsible for the very existence of matter as we know it.

Particles, fields, and symmetry: These are the great themes of twentieth-century physics. At the same time that it answers many questions, the Standard Model raises many new ones. Why do quarks come in six different "flavors"? Why are electrons so much lighter than the quarks, and neutrinos so much lighter than electrons? What's the deal with that remaining particle of the Standard Model, the Higgs particle, which hasn't been detected (yet)? What about dark matter and dark energy—where do they fit in? Perhaps the answer lies with new particles, or with new symmetries. Perhaps a completely new approach is needed. We will learn the current ideas and peek at what lies ahead for physics.

If all that had been accomplished in the past century were that a hundred or so fundamental atoms had been replaced by seventeen fundamental subatomic particles, it would still have been a great simplification in our understanding of matter. However, the Standard

Model goes much further. With a handful of additional parameters, it specifies *all of the interactions* between the particles. Including the parameters needed to specify the properties of the seventeen particles, there are just eighteen numbers needed to specify the Standard Model. Instead of an infinite number of possible groupings of atoms into molecules, and therefore an infinite number of chemical reactions whose rates must be measured, we have a mere eighteen parameters. All but one of the particles have now been produced in accelerator experiments, and the values of most of the parameters have been measured. The Standard Model puts us much closer to a complete understanding of the fundamental processes of the universe.

For this reason, the Standard Model is the greatest accomplishment of twentieth-century science. All you need is to measure the values of the eighteen parameters, and you know everything there is to know about everything in the universe, always excepting gravity. In principle, you could deduce the laws of thermodynamics, of optics, of electricity and magnetism, of nuclear energy, from the Standard Model. You could go on to explain the functioning of a star, a microbe, a galaxy, a human being, on the basis of those eighteen numbers.

If this is true, why haven't we been deafened with the popping of champagne corks, the cries of triumph, and the collective sighs of physicists retiring with the knowledge of a job well done? Why, instead, do we hear mysterious rumors of supersymmetry, string theory, and ten-dimensional spacetimes? One answer is the obvious omission of gravity from the Standard Model. Clearly, the job isn't done if such a major piece of the puzzle is missing. One might think that we could just tack a theory of gravity onto the Standard Model, call the result the New Standard Model, and be done. Unfortunately, the longer physicists work to do this, the more impossible the task appears. Our best theory of gravity (Einstein's theory of general relativity) and our best Theory of Almost Everything (the Standard Model) describe the universe in fundamentally different ways. It is far from clear how, or even if, these different structures can be reconciled. A two-word description of the structure of general relativity is *curved spacetime*, and that's about all I'm going to say about it. The structure of the Standard Model is what the rest of this book is about.

There is another reason why physicists aren't content to rest on their laurels and call it quits with the Standard Model: eighteen parameters are still too many! Why six quarks, rather than three, or two, or one? The top quark only showed up when physicists built a huge particle accelerator designed specifically to look for it. Couldn't the world have gotten along without it? A famous physicist derided the Standard Model saying, "Give me eighteen parameters and I can design an elephant." We would like the world to be even simpler, even more symmetrical, at its root. Ideally, physicists would prefer a single entity (maybe a string?) instead of the seventeen particles, and one law with one, or maybe no, parameters to be measured. (The great physicist John Archibald Wheeler has suggested that the ultimate laws of the universe, when we at last discover them, will seem so clear and obvious that everyone will nod their heads and agree that the world couldn't be any other way.) All of the known particles would arise from this fundamental entity behaving in different ways, like different notes played on a bugle.

Finally, the Standard Model can't be the end of the story because it fails to account for several important phenomena that have been discovered recently. Neutrinos have mass, according to recent experiments, whereas in the Standard Model they are massless. As we will see, neutrino masses can be accommodated in the Standard Model, but only somewhat awkwardly. Then there is the "dark matter" that astronomers tell us makes up most of the mass of the universe. Any theory that misses the majority of the stuff in the universe can't be complete!

But I am getting ahead of the story. To understand the greatest scientific accomplishment of the twentieth century, we need to back up and discover what physics was like back in the nineteenth century.

Chapter 1

The First Unifications

> If, in some cataclysm, all of scientific knowledge were to be destroyed, and only one sentence passed on to the next generation of creatures, what statement would contain the most information in the fewest words? I believe it is the atomic hypothesis…that all things are made of atoms—little particles that move around in perpetual motion, attracting each other when they are a little distance apart, but repelling upon being squeezed into one another.
>
> —Richard Feynman, *The Feynman Lectures on Physics*

Find a rock. Hit it with a sledgehammer. Take the smallest piece and hit it again. How many times can you repeat this procedure? As the bits of rock get smaller and smaller, you will need new tools and techniques. A razor blade, say, to divide the fragment, and a microscope to see what you're doing. There are only two possibilities: either you can go on dividing the pieces forever, or else you can't. If you can't, there must be some smallest uncuttable piece.

Leucippus and his pupil Democritus, Greek philosophers of the fifth century B.C., proposed that the division process had to end somewhere. The smallest piece was termed the *atom*, meaning "uncuttable." The atomic hypothesis flew in the face of common sense and everyday experience. Can you show us one of these atoms? Leucippus's opponents asked. No, replied the atomists. They are too small to be seen, invisible as well as indivisible.

More than 2000 years later, a new version of the atomic hypothesis was taking hold among scientists. By the beginning of the nineteenth century, it was becoming clear that all objects are composed of many small particles. The new concept of an atom was quite different from the Greek atom. For the Greeks, geometry ruled: atoms were supposed to be distinguished by their shape, even if that shape couldn't be seen. The new atoms were distinguished instead by their weight and their chemical properties. By the end of the nineteenth century, it was clear that atoms were not the whole story. Rather, there are two kinds of stuff in the world: particles and fields. Everything that we can see and touch is made up of indivisible particles. These particles communicate with each other by way of invisible fields that permeate all of space the way air fills a room. Fields are not made of atoms; they have no smallest unit. The particles determine where the fields will be stronger or weaker, and the fields tell the particles how to move.

The discovery of quantum mechanics in the twentieth century would overturn the straightforward view of a universe full of particles and fields. Another half century would pass before quantum mechanics and special relativity were assimilated into elementary particle physics to produce the most robust and successful scientific theory ever, the Standard Model of Elementary Particles. This chapter will reveal how the concepts of particles and fields were developed in the nineteenth century into powerful tools that brought unity to the diversity of physical theory.

Physics is the study of fundamental processes, of how the universe works at its most basic level. What is everything made of, and how do those components interact? There has been much talk in recent years of a "Theory of Everything," with string theory the leading candidate. For the physicist, that would be the ultimate achievement: a coherent set of concepts and equations that describe all of the fundamental processes of nature. The search for unifying descriptions of natural phenomena has a long history. Physicists have always tried to do more with less, to find the most economical description of the phenomena. The current drive for unification is but the latest in a long series of simplifications.

In the nineteenth century, physics was divided into many subdisciplines.

- **Dynamics**—The laws of motion. A sliding hockey puck, a ball rolling down a hill, a collision between two billiard balls could all be analyzed using these laws. Together with Newton's law of universal gravitation, dynamics describes the motions of planets, moons, and comets.

- **Thermodynamics**—The laws of temperature and heat energy, as well as the behavior of solids, liquids, and gases in bulk: expansion and contraction, freezing, melting, and boiling.

- **Waves**—The study of oscillations of continuous media; vibrations of solids, water waves, sound waves in air.

- **Optics**—The study of light. How a rainbow forms, and why a ruler looks bent when you dip it into a fish tank.

- **Electricity**—Why do my socks stick together when I take them out of the dryer? Where does lightning come from? How does a battery work?

- **Magnetism**—Why does a compass always point north? Why does a magnet stick to the refrigerator door?

By the beginning of the twentieth century, these branches had been reduced to two. Because of the atomic hypothesis, thermodynamics and wave mechanics were swallowed up by dynamics. The theory of electromagnetic fields subsumed optics, electricity, and magnetism. All of physics, it seemed, could be explained in terms of particles (the atoms) and fields.

The strongest evidence for the atomic hypothesis came from chemistry rather than physics. The *law of definite proportions*, proposed in 1799 by the French chemist Joseph-Lois Proust, declared that chemicals combine in fixed ratios when forming compounds. A volume of oxygen, for instance, always combines with twice that volume of hydrogen to produce water. The explanation comes from the atomic hypothesis: if water is compounded of one oxygen atom and two hydrogen atoms (H_2O), then two parts hydrogen and one part oxygen will combine completely to form water, with nothing left over.

By the end of the nineteenth century, it was already becoming clear that these chemical atoms were not, in fact, indivisible. In 1899, J. J. Thomson announced that the process of ionization involves removing an electron from an atom, and therefore "essentially involves the splitting of the atom."[1] In the early twentieth century, atoms would be further subdivided. "Splitting the atom" acquired a different meaning, namely, breaking apart the atomic nucleus by removing some of the protons and neutrons of which it is composed. An atom composed of neutrons, protons, and electrons was of course no longer uncuttable, but by that time the term *atom* was firmly established—it was too late to change. The most basic constituents of matter, those bits that could not be decomposed in any way, came to be called *elementary (or fundamental) particles*.

How does the atomic hypothesis allow thermodynamics to be reduced to dynamics? Take as an example the ideal gas law. Physicists experimenting with gases in the eighteenth and nineteenth centuries discovered that as a gas was heated, the pressure it exerted on its container increased in direct proportion to the temperature. No explanation was known for this behavior; it was an experimental thermodynamic law.

Let's apply the atomic hypothesis: Consider the gas in the container to consist of many small "atoms" that are continually in motion, colliding into each other and into the walls of the container like children in a daycare center. Now heat the gas to give the gas molecules more energy, raising their average speed. The pressure on the container walls is the cumulative result of many molecules colliding with the walls. As the temperature goes up, the faster-moving molecules hit the walls more frequently and with more force, so the pressure goes up.

A mathematical analysis proves that when you average over the effects of a large number of molecular collisions, the resulting pressure on the wall is indeed proportional to the temperature of the gas. What was formerly an experimental observation has become a theorem of dynamics. The properties of the gas are seen to be a direct result of its underlying structure and composition.

Dreams of Fields

To get an idea of the nineteenth century understanding of a *field*, start with a simple question: How does a compass needle know which direction is north? The compass needle, isolated inside its case, is not touching or being touched by anything other than the case itself, yet no matter how you twist and turn the case, the needle always returns to north. Like a magician levitating a body, some power reaches in with ghostly fingers and turns the needle to the correct position. Giving it the label *magnetism* doesn't answer the fundamental question: how can one object influence another without physical contact?

Isaac Newton struggled with the same question when he put forth his law of universal gravitation in 1687. He realized that the fall of an apple was caused by the same force that holds the moon in orbit around the earth, namely the earth's gravity. But how could the earth reach across 400,000 kilometers of empty space to clutch at the moon?

> That gravity should be innate, inherent and essential to matter, so that one body may act upon another at a distance thro' a vacuum, without the mediation of anything else, by and through which their action and force may be conveyed from one to another, is to me so great an absurdity, that I believe no man who has in philosophical matters a competent faculty of thinking, can ever fall into it. Gravity must be caused by an agent acting constantly according to certain laws; but whether this agent be material or immaterial, I have left to the consideration of my readers.[2]

The solution of this problem of "action at a distance," as it was called, came 200 years later in the field concept.

Imagine starting a barbecue in your backyard. Soon neighbors start dropping by: "How's it going? Oh, having a barbecue? Got an extra burger?" There's no need to contact them directly to tell them about the cookout, the aroma of the food carries the message. A (magnetic or electric) field works in a similar manner. Objects that display electric or magnetic properties are said to have an *electric charge*. This charge produces a field, rather like the barbecue produces an aroma. The larger the charge, the larger the field. A distant object doesn't need to be told of the presence of the charge, it only needs to sniff out the field in its immediate neighborhood, just as your neighbors sniffed out your barbecue. Thus, we say that the Earth behaves like a magnetic "charge" and creates a magnetic field filling the space around it. A compass needle, which is also a magnet, sniffs out the magnetic field and points along it. The compass, whether near the Earth or thousands of kilometers out in space, doesn't need to know where the Earth is or what it is doing. The compass responds to whatever magnetic field it detects, whether that field is generated by a distant Earth or a nearby refrigerator magnet.

Physicists represent a field by arrows. A bar magnet, for instance, is surrounded by a magnetic field that looks something like this:

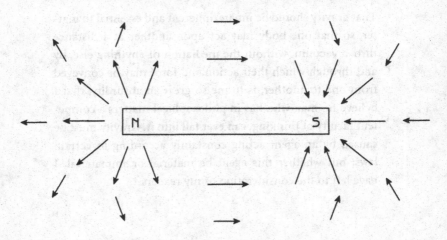

The stronger the field, the longer the arrow. Think of the magnetic field like a field of wheat: Each wheat stalk is an arrow, and the "field" is the entire collection of arrows. Unlike the wheat field, which only has a stalk every few feet or so, the magnetic field has an arrow at every spatial point. That is, to specify the magnetic field completely, one must give the strength of the field (length of the arrow) and direction of the field (direction of the arrow) *at every point in the entire universe.* Obviously, it would be impossible to experimentally determine the magnetic field at every point, even for a limited region, as it would require an infinite number of measurements. In real life, physicists must be content with having a pretty good idea of the field values in some limited region of space. To a physicist, the field is everywhere: in the air around you, penetrating the walls of your house, inside the wood of your chair, even inside your own body.

Around 600 B.C., the philosopher Thales of Miletos noticed that an amber rod rubbed with a silk cloth gained the power to attract small objects. We know this phenomenon as *static electricity.* (The word *electricity* comes from the Greek work for amber, *electron.*) You can perform Thales's experiment yourself. Tear some small bits off a piece of paper. Now rub a plastic comb on your shirt and hold the comb near the paper bits. If you are quick (and the humidity is low), you will see the paper jump and cling to the comb. This is a different force than magnetism: even a powerful magnet will not pick up the paper shreds, nor will the comb and the magnet exert a force on each other the way two magnets do. We call this new force the *electric force.* When you comb your hair and it stands out from your head, or when you take your clothes out of the dryer and they cling to each other, you are experiencing the electric force.

In all of these cases, there is a transfer of electric charge from one object to another. Benjamin Franklin discovered in 1747 that there are two types of electric charge, which he termed *positive* and *negative.* Ordinarily, objects like your socks have equal amounts of positive and negative charge, and so are electrically neutral (or uncharged).

Tumbling in the dryer, the socks pass negatively charged electrons back and forth like school children trading Pokemon cards. As a result, one sock ends up with excess negative charge and the other ends with excess positive charge. Opposites attract, according to the electric force law, so your socks cling together. When combing your hair, the comb strips electrons from your hair. Like charges repel, so your hairs try to get as far from one another as possible.

Electric interactions can be described in either a *force picture* or a *field picture*. In the force picture, we postulate a law of universal electricity (analogous to Newton's law of universal gravitation) that states, "every charged object in the Universe is attracted to (or repelled from, according to whether the charges are alike or opposite) every other charged object with a force proportional to the electric charge of both objects."

The field picture instead postulates a two-step process. In the first step, each charged object creates an electric field. (This is a different field from the magnetic field, but it can also be represented by drawing arrows at every point in space.) In the second step, each object feels a force proportional to the electric field at its location generated by all the other charged objects.

Mathematically speaking, there is one law to tell us what sort of field is produced by a given set of charges and another law to describe the force on a charge due to the electric and magnetic fields at the location of that charge. The sock doesn't need to "know" the location of every other charged object in the universe; it only needs to "know" the electric field at the sock's current location. In the field picture, objects respond to the conditions in their immediate surroundings rather than to the positions and movements of distant objects.

This may seem like a cheat: If both the force and field concepts give the same result, aren't they really saying the same thing in different words? Haven't we just hidden the "magic" action-at-a-distance behind an equally magical electric field? Indeed, it seems that the question of "How does the object know what distant objects are doing?" has merely been replaced with that of "How does the electric field know what distant objects are doing?"

To see the full power of the field concept, change the question from "How?" to "When?" Suppose one of your two electrically charged socks is suddenly moved to a new position: When does the other sock learn of the new circumstances? In the force picture, each sock responds to the current location of the other, so the force on one must change to a new direction as soon as the other is moved. In the field picture, however, we can imagine the possibility of a time lag between the movement of the sock and the change of the distant field. For locations near the new position of the sock that was moved, the field is centered at that new position, but for locations far away, the field is still centered at the original position of the sock.

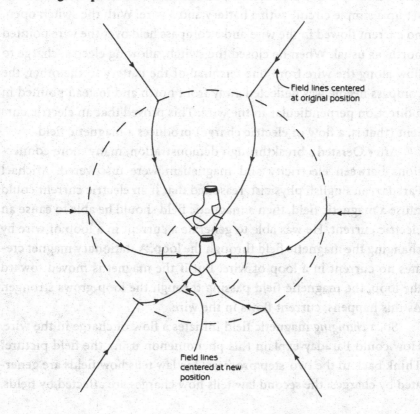

Field lines centered
at original position

Field lines
centered at new
position

If there is a time lag, there must be kinks in the field between the two regions. Perhaps as time goes on, the kinks move outward and the inner region that "knows about" the new location of the sock grows larger and larger. Can the theory of the electric field be modified to turn this "perhaps" into a definite prediction? To find the answer, we first need to find the connection between the two types of field: electric and magnetic.

The Marriage of Electricity and Magnetism

The first proof of a connection between electricity and magnetism was discovered by a Danish physicist, Hans Christian Oersted, in 1820. He set up a simple circuit with a battery and a wire. With the switch open, no current flowed in the wire and a compass held over the wire pointed north, as usual. When he closed the switch, allowing electric charge to flow along the wire from one terminal of the battery to the other, the compass needle was deflected away from north and instead pointed in a direction perpendicular to the wire. This proved that an electric current (that is, a flow of electric charge) produces a magnetic field.

After Oersted's breakthrough demonstration, many more connections between electricity and magnetism were discovered. Michael Faraday, an English physicist, reasoned that if an electric current could cause a magnetic field, then a magnetic field should be able to cause an electric current. He was able to generate a current in a loop of wire by changing the magnetic field through the loop. A stationary magnet creates no current in a loop of wire. But if the magnet is moved toward the loop, the magnetic field passing through the loop grows stronger. As this happens, current flows in the wire.

So, a *changing* magnetic field initiates a flow of charge in the wire. How could Faraday explain this phenomenon using the field picture? Think back to the two-step process: One law tells how fields are generated by charges; the second law tells how charges are affected by fields.

In Faraday's time, the second step was described by the Lorentz force law. According to this law, only an electric field can speed up or slow down a charge. A magnetic field can only change the direction of a charge that is already moving. Before the magnet starts moving, the electrons in the wire are stationary; the current meter indicates zero. Why, then, do the electrons in the wire begin to move when the magnet moves? Maybe the Lorentz force law is wrong, or maybe a moving magnet produces a brand new kind of force. Faraday, however, had a simpler explanation. If a moving charge could produce a magnetic field in Oersted's experiment, it seemed reasonable that a moving magnet could produce an electric field. It is this electric field that makes the current flow in the wire. Faraday took his experiment as proof that *a changing magnetic field creates an electric field.*

It was a Scotsman named James Clerk Maxwell who, in 1865, took the field concept (invented by Faraday) and gave it a clear mathematical formulation, incorporating the electric and magnetic force laws into a set of four equations, now known as Maxwell's equations. In developing these equations, Maxwell realized there would be an inconsistency unless it was possible for a changing electric field to generate a magnetic field. When Maxwell included this crucial modification in his equations for the electric and magnetic fields, he suddenly realized that not only all electric and magnetic phenomena, but also all the discoveries in optics, could be explained by his four equations, together with the Lorentz force law.

To understand the connection with optics, recall the kinks in the electric field that form when a charge is suddenly moved. As the charge moves, the electric field in its immediate area changes. We know from Maxwell's discovery that a changing electric field gives rise to a magnetic field, so the charge is now surrounded by both an electric and a magnetic field. Before the charge was moved, there was no magnetic field; in other words, there has been a change of magnetic field as well. According to Faraday, a changing magnetic field produces an electric field. A self-sustaining process arises, in which a changing electric field gives rise to a changing magnetic field, which in turn generates an additional electric field, and so on. The two effects reinforce each other, carrying the kinks in the field ever outward. The region near the charge that "knows about" the new location of the charge grows as the kinks move away from the charge.

This self-sustaining combination of changing electric and magnetic fields is called an *electromagnetic wave*. Maxwell found that the speed of these waves was related in a simple way to the two constants appearing in his equations. The numerical values of these constants were known from experiments that measured the strength of the electric and magnetic fields. Maxwell used the known values to find the speed of his electromagnetic waves, and discovered that they move at the speed of light. This could not be a coincidence: ordinary visible light must *be* an electromagnetic wave. The connection between light and electromagnetism has since been confirmed in many experiments.

A successful theory should not only explain phenomena that have already been observed and provide a framework for understanding them; it should also predict new phenomena. Experiments can then be designed to look for the new phenomena and test the theory. If Maxwell was right about light being a type of electromagnetic wave, there should be other forms of "light"—electromagnetic waves with a wavelength greater or less than that of visible light. There was nothing in his equations to prevent such waves; all that was necessary to produce them was to find some method of wiggling electric charges at the correct rate. The German physicist Heinrich Hertz set out to look for such things. He charged two metal balls that were separated by a small space. When the charge got high enough, a spark would jump across the space, carrying the negative charge over to the positively charged ball. The sudden movement of charge from one ball to the other created a kink in the electric field: an electromagnetic wave, according to Maxwell. On the other side of the laboratory, he set a loop of wire that had a tiny air gap in it. He knew that the wave should travel across the room at the speed of the light. When the wave hit the loop of the wire, it should cause an electric current to flow in the wire. Because of the air gap, this could only happen if a spark jumped across the gap. With the laboratory darkened, Hertz peered at the air gap and waited as the charge on the two balls built up. Whenever the spark jumped between the two balls, Hertz, on the other side of the room, saw a second tiny spark jump across the air gap in the wire loop.

Hertz found that his waves had a wavelength of about two feet, which is a million times longer than the wavelength of light. Electromagnetic waves with wavelengths of this size are now known as *radio waves*. Hertz's "broadcast," although not as gripping or informative as the Home Shopping Network, was nevertheless a tremendous accomplishment: the first radio transmission. The experiment provided direct proof that an electromagnetic wave was able to cross a room without the aid of wires.

It was later discovered how to produce electromagnetic waves with wavelengths between those of radio and light, and these were named *microwaves* and *infrared* radiation. Shorter wavelengths could be produced too, giving us ultraviolet radiation, X rays, and gamma rays. Today's society would not function without our knowledge of Maxwell's equations: We use radio waves for radio and TV reception, microwaves for microwave ovens and cellular phone links, infrared for heat lamps, ultraviolet for tanning booths and black light discos, X rays for medicine, and gamma rays for food decontamination. Visible light, running from red (the longest visible wavelength) to violet (the shortest), is only a small fraction of the electromagnetic spectrum.

Most of the electromagnetic "rainbow" is invisible to us. We can "see" ultraviolet waves in a vague, indistinct way, but not with our eyes: Our skin detects them and reacts with sunburn. The higher-energy X and gamma rays penetrate further into the body and can cause cell damage to internal organs. Mostly, though, we need to use specialized instruments as artificial eyes to expose the wavelengths we can't see directly. A radio or cell phone receiver uses an antenna and an electronic circuit, a dentist's x-ray machine uses photographic film to transform these signals into a form our senses can handle. Although generated and detected in a variety of ways, these waves are all fundamentally the same—traveling, self-sustaining electric and magnetic fields.

Thanks to Maxwell, electric and magnetic fields are much more than the cheat they might have at first seemed. The fields are not merely another way to talk about forces between particles. Electric and magnetic fields can combine to form electromagnetic waves that carry energy and information across great distances. Radio waves carry a signal from the station to your receiver, where they are decoded into news, music, and advertisements, without which your life would be incomplete. Light from the Sun traverses millions of miles of empty space; without this light, there wouldn't be life at all. Fields really exist and are a vital part of the world around us.

By the end of the nineteenth century, physicists had a clear picture of the basic physical interactions. According to this picture, everything in the universe is made of particles that interact by way of fields. Particles produce and respond to fields according to definite mathematical laws. The crowning achievements of physics were the two great unifications: the kinetic theory of thermodynamics, based on the atomic model, and Maxwell's electromagnetic field theory. These theories not only brought together many diverse phenomena, they made predictions about new phenomena and led to new experiments, new techniques, and new technology. The combined picture was so successful and so compelling that some physicists thought there was little left to do. Albert Michelson, a leading American physicist, said this in 1894:

> It seems probable that most of the grand underlying principles have been firmly established and that further advances are to be sought chiefly in the rigorous application of these principles to all phenomena which come under our notice... The future truths of Physics are to be looked for in the sixth place of decimals.[3]

The timing of such a pronouncement could hardly have been worse. By the end of the century, new phenomena were being discovered that were, to say the least, puzzling when considered according to the known laws of physics. In fact, two revolutions were about to occur in physics, and when the dust settled, the field and particle concepts would both be altered beyond recognition.

Chapter 2

Einstein's Relativity and Noether's Theorem

How deep is time? How far down into the life of matter do
we have to go before we understand what time is?
—Don DeLillo, *Underworld*

Take a peek into an alternate universe:

"Good morning, ma'am, Yellow Cab Company."

"Of course we can take you to the train station. How far
from it do you live?"

"Thirty miles? We'll need to leave at least an hour early.
Can't drive over 30 miles per hour—it's the law, you
know."

"Well yes, if you bring your laptop you can get some work
done on the way. But I wouldn't worry about that too
much if I were you. The trip only takes 10 minutes."

"No, no! We need to leave your house more than an hour
before the train's departure. Thirty miles at almost 30 miles
an hour means we need more than an hour to get there.
But your ride in the cab will only take 10 minutes. It's the
Time Effect, you know. The cab travels at 29.6 miles per
hour. That's pretty close to the speed limit."

"May I ask where you're going? Washington to Los Angeles!
That's a long trip. If you leave on Sunday you'll arrive

Thursday. You'll want your laptop—the train ride is 14 hours."

"Sure, you'd have a shorter trip if you went by plane. That's only a one-hour trip. The plane goes faster, nearly 29.998 miles per hour."

"No, you'll still arrive on Thursday. The plane only saves you travel time, not ground time. Of course, you pay through the nose for it."

"Right, then. Train it is. We'll pick you up on Sunday. Have a good trip!"

What is going on here? The speed of light in this alternate universe is just 30 miles per hour. According to special relativity, nothing can travel faster than the speed of light. ("It's not just a good idea, it's the law.") So, a trip of 30 miles will always take at least an hour, and a trip of 2,500 miles (Washington, DC to Los Angeles) will take at least three and a half days, in "ground time" as the cab dispatcher calls it. However, special relativity also tells us that there is a *Time Effect*: time runs at different rates depending on your state of motion. The effect gets stronger as you travel closer to the speed limit. The 83-hour (ground time) trip from DC to LA only takes 14 hours for train passengers (traveling at 29.6 miles per hour), and only one hour for plane passengers (at 29.998 miles per hour).

The strange effects of special relativity described in the alternate universe actually occur in our universe. However, these relativistic effects only become significant at speeds close to the speed of light. The reason we don't hear conversations like the one above is that the speed of light in our universe is 300,000 kilometers per second (186,000 miles per second), instead of 30 miles per hour, so the Time Effect is normally much too small to notice.

Let's return to our own world. Picture a flight attendant on an airplane traveling with a constant velocity in level flight. A bag of peanuts slips out of her fingers and falls to the floor. Now, intuition might lead us to think that, since the plane is moving forward during the time the peanut bag is falling, the bag will land toward the back of the plane from where the flight attendant is standing. This isn't what happens, however. The peanut bag lands right at her feet, exactly as if she had been standing on the ground.

In fact, life aboard a plane is remarkably unremarkable: coffee pours the same way it does on the ground, electrical devices function normally, voices sound the same. It is only when air turbulence suddenly lifts or drops the plane that cookies fly off trays and coffee leaps out of cups. As long as the plane is in uniform level motion, *no experiment performed on the plane will reveal that it is in motion.* Only by looking at an outside reference point (the ground, say) can it be determined that the plane is moving. We can summarize this in the principle of relativity:

> All steady motion is relative and cannot be detected without reference to an outside point.

Galileo propounded this principle already in the seventeenth century, when he noticed that if you dropped an object from the mast of a moving ship it landed at the base of the mast, in the same position relative to the ship as it would land if the ship was at rest.

The laws of physics tell us about the locations of objects at a particular time, and those locations must be measured from some reference point. The location of the falling peanut bag, for instance, might be specified as "20 meters from the back wall of the passenger compartment and 1 meter above the floor." We also need to choose a reference time, some event that tells us when to start the stopwatch. Such a choice of reference points in space and time is called a *frame of reference.* A frame of reference allows physics to become mathematics: Instead of a vague description—for instance, "the ball falls to the floor" —we can say *where* the ball is and *when* it is there. The numbers that specify the where and the when are meaningful because we have defined the reference points with respect to which they are measured.

The plane's frame of reference is in uniform motion with respect to the ground, as long as the plane is in straight, level flight at constant speed. Now, suppose some law of physics is different when in motion; the law of fluid flow, say, so that coffee pours differently when the plane is in motion. Then, all we would have to do to find out if we are moving would be to test that law by pouring a cup of coffee. If it pours as in a stationary frame, we must be at rest; if it pours as in a moving frame, we must be in motion. We would then have a way of detecting motion without reference to any outside point. Thus, another way to state the principle of relativity is this:

> The laws of physics are the same in any uniformly moving
> frame of reference.

A Relatively Moving Experience

As a high school student in the 1890s, Albert Einstein wondered about relative motion. What would happen, he asked himself, if you traveled fast enough to catch up with a light beam? What would it look like? Was there such a thing as light that stayed in one place?

Ten years later, Einstein had finished a degree in physics and had taken a position as a patent examiner third class in Bern, Switzerland. During his studies, he had learned Maxwell's equations, which purported to encapsulate everything there was to know about light. What would Maxwell's equations say about Einstein's high school conundrum? To his astonishment, Einstein found that it was simply impossible, according to Maxwell's equations, to move at the same speed as a light beam. This was a shocking discovery. Nothing in Newton's laws of motion suggested the possibility of an ultimate speed limit. Could the 200-year-old laws of motion be in error?

Other physicists who were aware of the dilemma assumed that it was the recently discovered electromagnetic equations that were incorrect. Einstein, younger and impatient with authority, assumed the opposite. What if Maxwell's equations were correct and Newton's laws of motion were wrong? Once the audacious first step was taken, the logical consequences could be derived using high school algebra. In a paper published in 1905 titled "On the electrodynamics of moving bodies," Einstein laid the foundations of a new dynamics, replacing Newton's laws of motion with the laws now known as special relativity.

Einstein's bold step was to add a new postulate to go along with the principle of relativity.

Einstein's Postulate: The speed of light is the same in all reference frames.

All of the weirdness of the Time Effect stems from this deceptively simple postulate. To see how, let's return to the airplane. Suppose a second flight attendant standing at the rear of the passenger compartment tosses a new bag of peanuts to the first attendant, who is 20 meters away. Let's suppose the bag was in the air for one second. From the reference frame of the plane, the speed of the bag was 20 meters per second.

However, things look different from the ground. During the time the peanut bag was in flight, the plane moved forward some distance, say 200 meters.

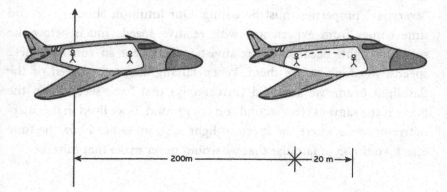

So, the bag went 200 meters + 20 meters = 220 meters, as viewed from the ground. Since the bag was in flight for one second, we conclude that its speed with respect to the ground was 220 meters per second. In other words, velocities add: The speed of the bag with respect to the ground is the sum of its speed as measured with respect to the plane and the plane's speed with respect to the ground.

We would expect the same thing to happen if the flight attendant had a flashlight instead of a bag of peanuts. Light travels at 300,000 kilometers per second, so one second after the flashlight is turned on, the forward edge of the beam will be 300,000 kilometers from the attendant. (We have to allow the flashlight beam to pass through the windshield of the airplane, or else assume we have a very long airplane!) Viewed from the ground, though, the beam will have traveled farther, just as the bag of peanuts traveled farther in the previous example.

In the reference frame of the ground, the beam, it seems, has traveled (300,000 kilometers + 200 meters) in one second. The beam moves a larger distance, but in the same amount of time: The speed of the beam is larger in the reference frame of the ground than in the reference frame of the plane. The addition-of-velocities rule clearly contradicts Einstein's postulate: If Einstein is right, the addition-of-velocities rule must be wrong. But we came to the addition-of-velocities rule considering only the everyday properties of distance and time. If the addition-of-velocities rule is wrong, our intuition about those "everyday" properties must be wrong. Our intuition about space and time comes from experiences with relative speeds much below the speed of light. Because the relativistic effects are so small at these speeds, we don't notice them. When talking about the speed of the flashlight beam, we assumed (incorrectly) that "one second" on the plane is the same as "one second" on the ground. If we lived in the alternative universe where the speed of light is 30 miles per hour, the time effect would be so familiar that we would never make that mistake.

If the speed of light is to be the same in both reference frames, it must be the case that either distance or time is measured differently in the two frames, or, perhaps, both. The answer, Einstein discovered, was "both." Neither time nor distance is absolute; they both depend on the relative motion of the observer. Space and time are thus inextricably intertwined. If you alter your rate of movement through space, you also alter your movement through time. As Hermann Minkowski put it in one of the first public lectures on special relativity, "Henceforth space by itself, and time by itself, are doomed to fade away into mere shadows, and only a kind of union of the two will preserve an independent reality."[1] The union of the two is what we now call *spacetime*.

Time runs differently for different observers: This is the explanation of the cab dispatcher's apparently nonsensical comments in the story that began this chapter. The 30-mile trip to the airport takes only 10 minutes for the person in the cab, but you still need to leave an hour early. When the passenger gets out of the cab, her watch will show that 10 minutes have passed since the cab ride began. For everyone else, however, an hour will have passed. Similarly, the plane trip from Washington, DC, to Los Angeles takes only an hour, but on arrival she will find that three and a half days have passed since she got on the plane.

Even more mind-boggling is that relativity requires that the situation be symmetric. If someone on the ground looked in through the windows of the train as it passed they would see the passengers moving in slow motion, breathing, talking, and eating six times slower than normal. The passengers in the train, on the other hand, look out of the window at people on the ground and see them in motion relative to the train, so those on the ground are seen as moving in slow motion.

Here, we have an apparent paradox: If each reference frame sees the other as slowed down, whose clock will be ahead when the passengers leave the train? The resolution of the paradox comes from the fact that the train must slow down and come to a stop in order for the passengers to disembark and compare watches with those on the ground.

When slowing down, the train is no longer in uniform motion, so the situation is no longer symmetric. It took Einstein 10 more years to extend his theory to cover nonuniform motion. The result was general relativity, a theory that dealt with cases of nonuniform motion, and incorporated gravity as well.

In the real world, the speed of light is not 30 miles per hour but 186,000 miles per second. The Time Effect only becomes large for speeds very close to the speed of light. Even the space shuttle moves at only a tiny fraction of light speed, so we do not notice these effects in everyday life. For elementary particle physicists, who accelerate particles to 99.999995 percent of the speed of light, the effect is enormous. These fast-moving but short-lived particles survive 3,000 times longer when in motion than at rest. If you could travel at that speed in a spaceship for a year, when you returned to earth 3,000 years would have passed. For the particles (and the physicists who study them), there is no *if*. The effect happens just as Einstein predicted. The Time Effect isn't science fiction or wild speculation; it is science fact.

The World's Most Famous Equation

Plant an acorn, and watch as it grows year by year to a tall tree. The tree's bulk obviously wasn't contained in the acorn. Where, then, did it come from? Some early scientists thought there must be a vital force associated with living things that created matter from nothing. Later, careful experiments showed that this was not the case. If you kept track of the water added, the weight of the soil, and, especially, the gases absorbed by the tree from the air, the mass of the tree was completely accounted for. The tree grows, not from nothing, but literally from thin air.

By the nineteenth century, the principle of conservation of mass was well established: Mass can neither be created nor destroyed. Technically, mass is a measure of the inertia of an object: how much it resists being pushed. More loosely, we can associate mass with the weight of an object. Think of a sealed box from which nothing can escape. Put anything you like inside the box before sealing it: a chemistry experiment, a potted plant with a battery-powered grow light, a pair of gerbils with enough air, food, and water for a lifetime. The conservation of mass implies that no matter what physical processes or chemical reactions are going on inside the box, no matter what creatures are being born there, growing, or dying, the box will always weigh the same.

Energy, according to nineteenth century physics, is a completely different beast. An object's energy depends on both its velocity and its mass. A bullet thrown by hand won't do any damage, but the same bullet projected at high speed from a gun can be lethal. A loaded dump truck that crashes when traveling 60 miles per hour will cause a worse wreck than a compact car traveling the same speed. An electromagnetic wave also carries energy, even though it is not made of particles—it is "pure" energy. In 1847, Hermann von Helmholtz proposed the law of conservation of energy: Energy can neither be created nor destroyed; it can only be converted from one form to another. For instance, sunlight shining into a car carries electromagnetic energy that is absorbed by the car seat and converted to heat energy.

In special relativity, energy and mass are no longer independent concepts. Einstein considered an object that emits electromagnetic waves. From the special relativity postulates, Einstein deduced that the object loses an amount of mass equal to the energy of the wave (E) divided by the speed of light (c) squared, *(old mass) – (new mass)* $= E/c^2$. He concluded that mass is really another form of energy. If the object could continue radiating energy until all its mass is gone, it would release an amount of electromagnetic energy equal to $E = mc^2$. The speed of light is very large: $c = 300,000$ kilometers per second, so a tiny amount of mass produces a large amount of energy. A grain of salt, if all its mass could be converted to energy, could power a light bulb for a year.

To put it another way, suppose you had a microwave oven that didn't just heat the food, it actually created it out of electrical energy. No need to put anything into the oven, just spin the dial to Hamburger, press Start, and out pops a steaming quarter-pounder. Sound enticing? But it would take about three billion kilowatt-hours of electricity, at a cost of about a hundred million dollars. Suddenly McDonald's is looking pretty good.

Even Einstein had misgivings about overturning the time-honored conservation laws of mass and energy. In a letter to a friend, he wondered "if the dear Lord ... has been leading me around by the nose"[2] about mass-energy equivalence. Today, the conversion of mass into energy is a matter of course: Nuclear power plants operate on this principle. Perhaps the most dramatic demonstration occurred when the first atomic bomb was exploded in New Mexico on July 16, 1945, converting a raisin-sized amount of mass into energy.

The equation $E = mc^2$ is valid for an object at rest. For an object moving with speed v, Einstein derived a different formula:

$$E = \frac{mc^2}{\sqrt{1 - \dfrac{v^2}{c^2}}}$$

According to this equation, as the object's speed approaches the speed of light, its energy grows to infinity. An infinite amount of energy is impossible to achieve, therefore nothing having mass can ever reach the speed of light. It's like a race in which the contestants move halfway to the finish line each time a whistle blows: They never reach the finish because they always have half of the distance remaining. Similarly, each time energy is added to an object the increase in speed is less. When the object is moving at a speed very close to c, it takes an immense amount of energy to increase the speed by even a small amount. Particle physicists spend vast amounts of the taxpayers' money to build huge machines to nudge the particles a bit closer to the speed of light. As we will see, it is not the tiny increase in speed that interests these researchers, but the great gain in energy that accompanies it.

A *massless* particle, however, can travel at the speed of light; in fact, it *must* do so. If such a particle could exist, it would carry energy, but it could never be brought to rest and weighed. For this reason, physicists say such particles have no *rest mass*. Because they have energy, it isn't strictly correct to call them massless. A box full of such particles zipping back and forth would weigh (very slightly) more than the same box when empty. These massless particles will reappear in later chapters; keep in mind, though, that we're using the word "massless" to mean particles that have no rest mass, carry energy, and always move at the speed of light.

The idea that the amount of "stuff" in the universe doesn't change—that is, that mass is conserved—makes intuitive sense: You can saw up a log into boards, but the total weight of the boards, plus the weight of the splinters, shreds of bark, and sawdust left over from the sawing must be the same as the weight of the original log. Energy is a much more abstract idea. A fast-moving object has more energy than the same object when moving slowly. Only by making careful measurements and combining the measured values in the correct mathematical relationship do we discover that the particular combination we call *energy* has the same value at all times. Einstein's discovery of the equivalence of mass and energy reveals that energy is just as fundamental as mass; energy counts as part of the "stuff" of the universe, too. What Helmholtz's principle of energy conservation had hinted at, special relativity made indisputable. Energy is not just a mathematical tool; it is a fundamental physical entity.

In addition to the equivalence of mass and energy, the space/time connection in special relativity also has deep philosophical consequences. Physical facts have meaning only insofar as they pertain to a particular observer. If Albert and Betty clap nearly simultaneously, one observer may report that Albert clapped first, whereas a second observer, in motion with respect to the first, may report that Betty clapped first. It makes no sense to ask, "Who *really* clapped first?" The question assumes that one viewpoint, one reference frame, is valid or "real" and the other is not. But time is not absolute; it is a property of a particular frame of reference. Both observers' viewpoints are equally valid. Do not be confused by the term *viewpoint* into thinking that the difference is merely a matter of opinion. A viewpoint here has the very specific meaning of a frame of reference, a choice of reference points in space and in time from which all measurements are made. We are talking about differences of *measurement*, not of opinion. Moreover, an observer who understands special relativity can easily change viewpoints, converting all his measurements into the reference frame of the other person. Doing so allows him to understand the other's conclusion about the order in which the events occurred.

Special relativity taught physicists that only things that can be measured have meaning: There is no way to measure which event really happened first, so the question is meaningless. It was the beginning of a fundamental shift of philosophy in science, from asking questions of what *is* to asking what *can be known*. This shift would become even more prominent in the rise of quantum mechanics.

A Radically Conservative Idea

There is a deeper meaning behind the conservation of mass-energy: It is a necessary consequence of a fundamental symmetry of the physical universe. The connection between symmetries of nature and conservation laws was discovered by a young German mathematician, Amalie Emmy Noether, who had been working with the great mathematician David Hilbert on Albert Einstein's new theory of gravity, the general theory of relativity.

Noether was forced to struggle against the institutional sexism of her time. Hilbert tried to get her a paid position at the University of Göttingen in 1915, but was turned down on the basis of "unmet legal requirements," a roundabout way of saying "we don't hire women professors." At a faculty meeting, Hilbert replied, "I do not see that the sex of the candidate is an argument against her admission as *Privat-dozent* (Lecturer). After all, we are a university, not a bathing establishment." Unfortunately, his eloquence was ineffective. After several years as an unpaid lecturer at Göttingen, she was finally given a paid position in 1923 and was allowed to oversee doctoral dissertations. She became one of the founders of a new branch of mathematics known as *abstract algebra*. This is not the same algebra you learned in school. Ordinary high-school algebra deals with the properties of numbers and the rules for manipulating them. The algebraists noticed that other mathematical objects obey some of the same algebraic rules as ordinary numbers. By abstracting and generalizing the principles of algebra, abstract algebra pulled together results for numbers, vectors, matrices, and functions. Results proved in the general theory automatically applied to any system that obeyed the general rules. Abstract algebra is still a cornerstone of mathematics today.

Because of her Jewish background, Noether was forced to give up her position at Göttingen when the Nazis came into power. She moved to the United States to take a teaching position at Bryn Mawr College, a small women's college in Pennsylvania. Norbert Weiner wrote in support of her application, "Leaving all questions of sex aside, she is one of the ten or twelve leading mathematicians of the present generation in the entire world and has founded ... the Modern School of Algebraists."[3] Sadly, she died only two years later after a surgical procedure.

To the mathematician, Emmy Noether is the founder of a fundamentally important branch of mathematics and the author of many important theorems. To the physicist, there is one result of hers that is so important it is known among physicists simply as *Noether's theorem*. This was a minor part of her post-doctoral work that grew out of her work on general relativity. Noether's theorem (as I shall call it, too, since this is a book about physics, and bar my door against enraged mathematicians) relates the symmetries of a physical system to the conserved quantities, like energy, that can be found for the system.

We usually think of symmetry in terms of objects like snowflakes.

If someone were to rotate a perfect snowflake by 1/6 of a full circle (or 60°) when you are not looking, you would have no way of knowing it was rotated. The snowflake is said to be *invariant* under such a rotation. It is also invariant under rotations of 120°, 180°, 240°, 300°, and, of course, 360°.

The snowflake is also invariant under *mirror symmetry*: It looks just the same when viewed in a mirror. The human body very nearly possesses mirror symmetry; however, if you part your hair on one side rather than in the center you will look different in the mirror than in a photograph. Even if you attempt to part your hair exactly down the center, there are subtle asymmetries that will give the game away. The procedure of replacing something with its mirror image is called the parity operation; if the situation remains unchanged it is said to be invariant under parity.

It is easy to think of objects with other symmetries. The 12-sided dodecagon is invariant under rotations of one twelfth of a circle (30°), or any multiple of 30°. A perfect circle, on the other hand, is invariant under any rotation whatsoever. Because the circle can be continuously rotated, it is said to be invariant under a *continuous symmetry*. Rotations by discrete amounts, as for the snowflake and dodecagon, are called *discrete symmetries*. Parity (invariance) is another discrete symmetry.

Symmetry in physical systems carries a different meaning than simple geometrical invariance. Instead of asking whether an experiment looks identical when rotated (geometrical invariance), we ask whether the laws of physics are invariant. In other words, do objects behave in the same way when the system is rotated? A collision between two billiard balls doesn't have geometrical symmetry; it is easy to tell if the billiard table has been rotated. In physical terms, though, nothing important has changed. If the speed of the incoming balls and the angle between them is the same in both experiments, the speed of the outgoing balls will also be identical in the two experiments, as will the angle between them. It is in this sense that physicists talk of the *rotational invariance* of a physical system. Rotating the initial setup leads to a rotated, but otherwise identical, outcome. Rotational invariance is a continuous symmetry; the billiard table can be rotated by any amount without affecting the outcome.

Suppose we have a theory that we want to check for rotational invariance. We obviously can't solve the equations of the theory for every experiment that could conceivably be done, and then check that we get the same answer again when the experiment is rotated. Fortunately, this is not necessary. We can instead check the equations themselves for symmetry. The equations may involve directional quantities, those quantities that are represented by arrows (known as *vectors*). For instance, the Lorentz force law involves the particle's velocity, the electric field, and the magnetic field, all of which are vectors. Other quantities, such as the mass of a particle, have no direction associated with them and so are unchanged by a rotation. To check the equations for rotational invariance, we mathematically rotate all of the directional quantities, plug the rotated values back into the equations, and check if the resulting equations have the same form as the original equations. If they do, the theory is rotationally invariant. Maxwell's equations, the Lorentz force law, the laws of dynamics, indeed, all known laws of physics, are rotationally invariant.

Most physical systems are endowed with two additional continuous symmetries, which we can call the *space shift* and *time shift* symmetries. (Physicists call them *spatial translation invariance* and *time translation invariance*.) We expect that any experiment can be moved four feet to the left, or moved to New York, Helsinki, or Canberra, without affecting the outcome. Assuming, that is, that any purely local conditions, like the altitude, temperature, or local magnetic field, don't affect the outcome. A system that can be moved from place to place without affecting the outcome has space shift invariance. Similarly, we expect that it won't make a difference what time we start an experiment. Starting two hours later, or a week from next Thursday, will only change the timing of the subsequent events, not the ultimate outcome. Again, local effects must be excluded: If you have a date for tonight you'd better not try the experiment of showing up a week from next Thursday! Setting a time for the date creates a *local condition*: All times are no longer equivalent, so the situation no longer has time shift invariance. The laws of physics admit no privileged time: They have time shift invariance.

Here's where Noether's theorem comes in. The theorem declares that there is a *conserved quantity* associated with every continuous symmetry of a physical system. Previously in this chapter, we discovered that energy is a conserved quantity—it can't be created or destroyed, only converted from one form to another. Is there a corresponding symmetry? Yes, in fact, it is time shift invariance. Noether's theorem provides the connection: From the time invariance of the theory, the expression for the conserved quantity can be derived, and it turns out to be just what we call the energy.

The other two continuous symmetries we have encountered, space shift invariance and rotational invariance, naturally correspond to other conserved quantities. The conserved quantity corresponding to space shift invariance is momentum, the inertia of an object due to its forward motion. Conservation of momentum is what tells us that, in a head-on collision between a Mack truck and a VW Beetle, the combined wreckage will be traveling in the direction the Mack truck had been traveling before the collision. The conserved quantity corresponding to rotational invariance is called *angular momentum*, which is, roughly speaking, the amount of spin of an object. Conservation of angular momentum explains why an ice skater who is in a spin will rotate faster when he pulls his arms in. (It's fun to test this yourself using a swivel chair. Hold your arms out to the side and start the chair spinning, then pull in your arms. The effect is increased if you hold a heavy book in each hand.)

As an example of space shift invariance, think about a skateboarder in a half-pipe. Viewed from the side, the half-pipe doesn't have space shift invariance. If the pipe were suddenly shifted to the left, the skateboarder would find herself in midair. A space shift in this direction causes a change in the physics of the situation. Viewed lengthwise, however, the half-pipe *does* have space shift invariance.

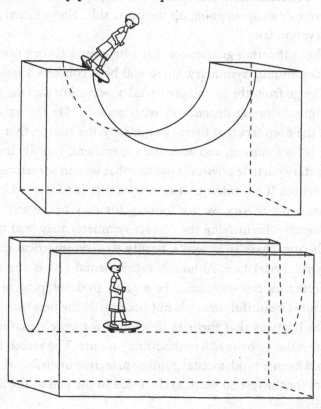

If the pipe were suddenly shifted lengthwise, the skateboarder would-n't notice that anything had changed. As long as she wasn't near either end of the pipe, she could complete her maneuver as if nothing had happened.

As a result, there is a conservation law for the lengthwise direction, but not for the crosswise direction. A skateboarder riding lengthwise down the pipe moves at constant velocity. Here, velocity is the conserved quantity. Actually, the conserved quantity for space shift symmetry is momentum, which is mass times velocity. In this case, the skateboarder's mass isn't changing, so constant momentum implies constant velocity. In the cross-pipe direction, though, momentum and velocity are not constant. The skateboarder speeds up as she descends the pipe and slows again going up the other side. No symmetry means no conservation law.

Noether's theorem guarantees that whenever a theory is invariant under a continuous symmetry, there will be a conserved quantity. It allows us to go from the seemingly trivial observation that the result of an experiment doesn't depend on what time of day the experiment begins to the deep fact that there is a quantity, the energy, that remains the same before, during, and after the experiment. Equally important for elementary particle physics is the fact that we can sometimes go the other direction: If we notice that some quantity is conserved in all our experiments, the theory we are looking for may be invariant under some symmetry. Identifying the correct symmetry may lead us to the correct theory. There is as well a beauty to a symmetrical theory, as with a symmetrical face. Although experimental test is the ultimate arbiter, aesthetics can sometimes be a guide in developing new theories. The most beautiful theory is not necessarily the best theory, but it sometimes happens that theories developed for purely mathematical reasons turn out to be useful in describing nature. The search for symmetries has been a fundamental guiding principle of elementary particle physics for the last 50 years, and has led in the end to the Standard Model.

Special relativity dramatically changed physicists' ideas about space and time, mass and energy. It left intact the concepts of particles and fields as the stuff of which things are made. The actors remained the same; only the stage on which they were acting was changed. Quantum mechanics, the other great conceptual development of early twentieth-century physics, would retain nineteenth-century ideas of space and time but would revolutionize ideas about particles and fields.

Chapter 3

The End of the World As We Know It

Can Nature possibly be as absurd as it seems to us in these atomic experiments?
—Werner Heisenberg

The crack of the bat. The center fielder, peering intently at the batter, hesitates only a fraction of a second, then turns and runs at full speed away from home plate. As he nears the home run wall he turns, raises his glove, and the baseball falls neatly into the pocket.

The center fielder has solved the quintessential problem of classical physics, which is to say nineteenth-century physics. During the brief interval after the ball was hit, he estimated its speed and direction. Years of experience allowed him to deduce where the ball would land and how fast he needed to run to be there in time to catch it. This is exactly what a physicist seeks to do: predict the future based on knowledge of the present. That is, knowing the *position* of the ball and the *velocity* (that is, the speed and direction) of the ball at the time it was hit, the physicist attempts to predict the position and velocity at any later time.

The physicist summarizes her years of experience in a set of mathematical laws. In the case of the baseball, those laws must encapsulate the effects of gravity and air resistance on the ball's motion. More generally, as we've seen, we need to know what fields there are and how those fields affect the object's motion. Given a snapshot of the universe, or some part of it, at any time, the laws of physics let us picture the world at any other time. This classical worldview consists of:

- A list of every object (particle or field) in the world and its location. For fields, which extend throughout space, we need the value of the field at each point in space, for example, the lengths and directions of the electric and magnetic field arrows.

- A description of how every object is changing. For particles, this means the velocity, which tells how the particle's position is changing. For fields, this means the rate of change of the length and direction of the field arrows at each point in space.

- A theory of interactions (forces) between particles and fields. We need to know how particles create fields and how particles respond to fields.

The first two items completely specify the state of the universe at a particular time. The third item allows us (in principle) to extend the description to another time, future or past. In other words, if we know the state of the universe at any one time, we can completely predict the future, and completely reconstruct the past.

Special relativity, counterintuitive as it is, still fits comfortably into the classical worldview. Predicting the state of the universe at any time, we know from special relativity, only makes sense with respect to a chosen frame of reference. Special relativity merely changes some of the equations we use to extrapolate our snapshot of the world from one time to another.

Quantum mechanics completely overturned the classical worldview. Quantum mechanics denies that such a classical description can ever, even in principle, be given. Not only are the future and past unknowable, in the sense of knowing the first two items in the preceding list, but even knowing the present (classical) state of the universe is impossible. Nor is it a problem with finding the information in an infinite number of places at once: Even for one particle and one instant in time, it is impossible in principle to know both the location and the velocity precisely.

Why did physicists abandon the predictive power of classical physics and embrace the probabilities and uncertainties of quantum mechanics? Some observers have suggested that they were infatuated with eastern mysticism and mathematical group theory and were looking for a way to work them into physics. On the contrary, they were forced by Nature to accept this strange description of her. Two phenomena in particular triggered the quantum revolution: the photoelectric effect and the structure of the atom.

If you have never read about quantum mechanics before (perhaps even if you have), you will no doubt find it confusing, maybe incomprehensible. If you do, take heart! As the brilliant (and Nobel prize-winning) physicist Richard Feynman put it:

> It is my task to convince you not to turn away because you don't understand it. You see, my physics students don't understand it either. That's because I don't understand it. Nobody does.[1]

The Unbearableness of Being Light

In the seventeenth century, opinion was divided: Did light consist of particles emitted by the light source and absorbed by the eye, or was it a wave, a vibration of some medium, the way sound is a vibration of the air? Isaac Newton began investigating light in 1665 using a prism he bought at a traveling fair. (His groundbreaking demonstration that the colors of the rainbow could be recombined into white light had to wait until the fair came around again and he could buy another prism.) Newton eventually became a strong proponent of the particle model. Using the particle model, he was able to explain several important optical phenomena.

First, light travels in a straight line. Because particles travel in a straight line unless acted on by an external force (a fact now known as Newton's first law of motion), Newton could explain the straight-line motion of light by making one additional assumption: that the light particles are weightless, that is, they are not affected by gravity.

Second, if you shine a flashlight at a mirror, the beam bounces off at the same angle as the incident angle. This behavior is also explained by the particle model; a ball bouncing off the floor or wall does the same thing.

Third, when light travels from one medium to another, a light beam changes direction. For instance, when light goes from the water in a fish tank into air, the light path is bent. This effect, known as *refraction*, makes a ruler partially immersed in water appear crooked. Newton was able to explain refraction by assuming that a force acts on the light particles when they are near the interface between the water and the air, so that their speed changes as they travel into a different medium.

At the same time, Christian Huygens, a Dutch physicist, was developing a theory in which light is a wave. He was able to explain reflection and refraction using this model, but the straight-line propagation of light was more difficult. The problem is that waves tend to bend around corners. This is why it is possible to hear someone talking who is in the next room, even if you can't see them. The sound bends around the corner, but the light doesn't.

The spreading of a wave as it passes a corner or an obstacle is called *diffraction*. In a report published posthumously in 1665, the Italian physicist Francesco Grimaldi demonstrated that light does, in fact, spread out as it passes through an opening in an opaque screen, forming a series of dark and light bands a small distance into the shadow region, as shown in the following image.

Grimaldi proved that light *does* bend around corners—it just doesn't bend nearly as much as sound does. The tiny amount of bending makes sense if the wavelength of light, the peak-to-peak distance

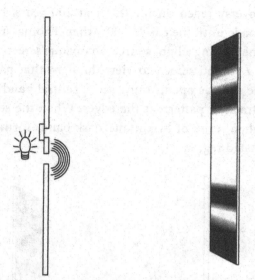

between successive waves, is very small. In fact, light must be graced with a wavelength a hundred thousand times shorter than that of sound to explain Grimaldi's observations.

You can check Grimaldi's results for yourself. Find a straight-edged object such as a ruler, a pen, or a Republican. Hold the ruler between your eye and a light source with a sharp outline, a fluorescent light fixture, for instance. (It helps if the room is dark except for the light source.) The light will seem to come *through* the ruler's edge. What you are actually seeing is light that has bent around the edge of the ruler.

Even when Newton came across Grimaldi's report of the diffraction of light by an aperture he stubbornly stuck to his particle model. He explained the spreading of the light by invoking a force on those light particles that just skim the edges of the opening. The bright and dark bands he explained by supposing that the force was sometimes attractive and sometimes repulsive, so that "the Rays of light in passing by the edges and sides of bodies (are) bent several times backwards and forwards, with a motion like that of an Eel." Not a very good explanation, but with Newton's magisterial reputation behind it, the particle model remained popular.

The controversy raged on for the next 200 years, but not much progress was made until the early 1800s when Thomas Young set up a simple experiment using a light source, an opaque screen with two narrow slits, and a second screen to view the light that passes through. When only one slit was open, Young saw a central band of light, with Grimaldi's diffraction pattern at the edges. When the second slit was opened, though, a series of bright and dark bands formed within the central illuminated region.

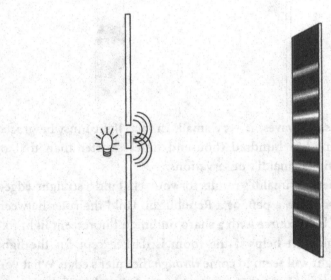

Now, imagine standing at the location of the viewing screen with your eyes at the position of one of the dark bands on the screen. Before the second slit is opened, you see bright light coming from the open slit. When the other slit is opened, you suddenly see no light at all! In other words, portions of the illuminated region become darker upon the addition of the light from the second slit. This cannot happen in Newton's theory, Eels or no Eels. Opening the second slit can only add light particles: You should get more light, not less. Young, using the wave theory of light, was able to explain how light from one slit could "interfere" with the light from the other slit.

To understand Young's explanation, let's look at water waves generated by wiggling two sticks at opposite ends of a tank. The wave on the left will travel to the right and the one on the right will travel left until the waves meet in the middle. The two sticks are being wiggled in opposite directions, so the front of the lefthand wave is *above* the normal surface of the water while the front of the right-hand wave is *below* the normal surface. Therefore, when the two waves meet at the midpoint of the tank, the two disturbances will cancel each other. The water level at the exact midpoint of the tank will always remain at the original, undisturbed level. The cancellation of two waves is called *destructive interference.*

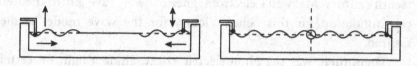

Destructive interference is the source of the bright and dark bands that appear when the second slit is opened in Young's two-slit experiment. Any point (other than the exact center point) on the viewing screen is slightly farther from one of the slits than from the other slit. There are certain points on the screen where the path difference is just enough so that the light wave from one slit is doing the exact opposite of the wave from the other slit. The extra distance is just enough for the wave from the lower slit to go through half of its cycle.

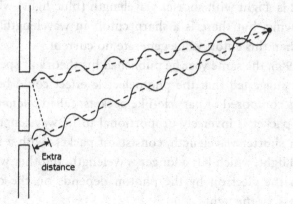

Extra
distance

As a result, it is down whenever the other wave is up, and vice versa. When the two waves meet at the viewing screen, they will always cancel. The dark bands, called *nodes*, are the points on the screen where this cancellation happens. Between the nodes are the points where the two waves add: the bright bands.

Destructive interference is the hallmark of wave phenomena. As we have seen, no particle model can explain how the brightness can decrease at the nodes when a second slit is opened. Young's experiment convinced physicists that light is in fact a wave. By the end of the nineteenth century, Maxwell's electromagnetic theory gave a firm theoretical foundation for this belief. Victory for the wave model seemed assured.

Then, there was the photoelectric effect. Shine a light on certain metals and an electric current will start to flow. Physicists studying this effect assumed that the energy in the light was being transferred to the electrons in the metal, but there was a puzzle. Blue light causes current to flow, but red light doesn't. The puzzling aspect is that, in Maxwell's theory of light, the energy provided by the light depends on the brightness of the light. So brighter light should cause more of a current regardless of what color, or wavelength, of light you are using. This is not what happens, however. Light of long wavelength (towards the red end of the spectrum) doesn't cause any current to flow, no matter how bright it is. Light with shorter wavelength (blue light) will start a current flowing, but there is a sharp cutoff in wavelength. Wavelengths longer than this cutoff value generate no current.

In 1905, the same year he published his theory of special relativity, Einstein suggested that the photoelectric effect could be explained if light was composed of particle-like packets, called *photons*. The energy of each packet is inversely proportional to its wavelength: Blue light, having a shorter wavelength, consists of packets with a higher energy than red light, which has a longer wavelength. In other words, the kick given to the electron by the photon depends on the color, not the brightness, of the light.

Picture the electrons in the metal as soccer balls sitting on a low-lying playing field surrounded by higher ground. It is Saturday morning and the kids on the soccer team, wearing bright red uniforms, are trying to kick the balls up the hill onto the higher ground. None of them are strong enough to get a ball up the hill, though; each ball goes up a bit and then rolls back down. No matter how many kids are on the field, none of the balls ever makes it up the hill. Then, Soccer Mom comes onto the field, wearing her light blue coaches' outfit. She easily boots the ball up the hill and it rolls off into the distance.

Einstein's proposal worked beautifully for the photoelectric effect. Long-wavelength (red) photons, like the kids on the soccer field, don't give the electron enough energy to make it up the "hill," which is formed for the electron by the force of the atoms in the metal. Brighter light has more photons and carries more energy, just as in Maxwell's theory. But the electron only absorbs one photon at a time, so no matter how many photons there are, none make it up the hill to become conduction electrons. As with the young soccer players, having more on the playing field doesn't help. Blue light has short-wavelength photons that carry more of a punch, like Soccer Mom. An electron that absorbs a photon of blue light gains enough energy to make it up the hill and become part of the flowing current.

For 40 years, in test after test, Maxwell's electromagnetic theory had triumphed, proving conclusively (so physicists thought) that light was a wave. An entire industry of the generation, distribution, and use of electrical power was built on this understanding of electricity and magnetism. Now Einstein was proposing that light had a particle nature, as Newton had suggested more than two centuries earlier! The old question, "Is light a particle or a wave?" was suddenly resurrected from the grave that Maxwell had made for it. How can it be that Maxwell's equations are so brilliantly successful in explaining wave phenomena like interference and diffraction if light is "really" a particle and not a wave at all? The answer only came with the development of relativistic quantum field theory, the subject of Chapter 5, "The

Bizarre Reality of QED." As a provisional solution, though, we can think of light as a field that indeed obeys Maxwell's equations, but at the same time it is a field that comes in chunks—it can only be emitted or absorbed by matter as a whole packet of the correct energy for its wavelength.

Einstein's proposal was not yet a replacement for Maxwell's theory; it was still incomplete. The concept of photons was difficult to square with Maxwell's electromagnetic wave theory of light. Almost a half-century would pass before a complete quantum theory of light was found, one that made sense of both Maxwell's equations and the photon nature of light.

The Death of Certainty

The photoelectric effect was a problem for classical physics, but it was easy to imagine that photoelectric metals had some peculiarity that caused the strange behavior. The structure of the atom raised difficulties that could not be shrugged off so easily. In 1911, Ernest Rutherford deduced from his experiments that atoms were composed of a positively charged core, called the nucleus, surrounded by much lighter negatively charged electrons. Thus, the atom was thought to look like a tiny solar system, with the nucleus as the sun, electrons as the planets, and the electrical attraction of the positively charged nucleus and the negatively charged electron replacing gravity as the force that holds it together.

According to Maxwell's theory, however, an electron in such an orbit would have to emit electromagnetic radiation, thereby losing energy, which would send it into a "death spiral," which could not end until the electron reached the nucleus. With all the negatively charged electrons in the nucleus canceling out the positive nuclear charge, there would be no electric repulsion keeping the nuclei at atomic distances from each other. In a fraction of a second a house would collapse to the size of a grain of sand. No objects would retain their shapes or size. All matter would be unstable. Here was a serious difficulty. It couldn't be brushed off as a peculiarity of a few special materials. The laws of physics were saying that matter as we know it simply can't exist. It was time for some new laws of physics.

Physicists didn't see the radiation from the death spiral, but they did see a different pattern of electromagnetic radiation coming from atoms. Since the mid-nineteenth century, it was known that each chemical element has a characteristic spectrum of light. Send the light from an incandescent bulb through a prism and it spreads into a complete rainbow. If you use a fluorescent bulb, however, you see instead a series of bright lines of color. A fluorescent bulb is filled with a particular gas, and each different gas has its own line spectrum, a chemical fingerprint. The origin of these bright lines was a complete mystery. Quantum mechanics, the theory that was developed to explain them, did more than change a few of the equations of classical physics. It required a completely new view of reality and of the goals of physics.

It took many years and the combined effort of many brilliant minds to create the theory of quantum mechanics. Rather than lead you through the historical development, I am going to jump to the end and describe the picture of the microworld that physicists ended up with. Discussion of quantum mechanics by itself could occupy an entire book; indeed some excellent books have been written on the subject. See the "Further Reading" section for some suggestions.

In classical physics, the universe is composed of particles and fields. A complete description of the world at any instant must specify the locations of the particles, the values of the fields, and how both are changing. From this information and the laws of interaction between particles and fields, the complete future of the universe can be predicted.

In quantum mechanics, the basic picture is radically different:

A. The motion of any particle is described by a wave, known as the *wavefunction* or *quantum field*.

B. The probability for the particle to be detected at a given point is the square of the quantum field at that point.

C. The quantum field changes according to a mathematical law known as the Schrödinger equation.

In (A), we have the first hint that particles and fields are not such wildly different entities as they appear in classical physics. The full realization of this wave-particle duality comes in relativistic quantum field theory, where particles and fields are treated identically. Relativistic quantum field theory is, then, a unification of particles and the forces acting on them. The terrible price we have to pay for the unification of particles and fields is revealed in (B). The laws of quantum mechanics are *random*; only probabilities can be determined. We give up the ability to predict the future that classical mechanics promised us—we can only have limited, probabilistic knowledge of the outcome of any experiment. Not only that, but even the present is not completely knowable. In quantum mechanics, a complete description of the current state of the universe consists of specifying the value of each quantum field at every point in space. From (B), we learn that such a description doesn't tell us where the particle is, it only gives the probability of finding a particle at any given location. Moreover, the wave description of particles involves a tradeoff: the more precisely the location of a particle is known, the more uncertain is its velocity. This result is called the Heisenberg uncertainty principle.

The idea that matter had wave-like properties, a key aspect of quantum mechanics, was first proposed in 1923 in the doctoral thesis of a young French nobleman, Luis-Victor Pierre-Raymond de Broglie. De Broglie was inspired by Einstein's suggestion that light might have particle-like properties, even though the interference effects "proved" it was a wave. If a wave could act at times like a particle, de Broglie reasoned, why couldn't a particle, the electron for instance, act like a wave? According to de Broglie, every particle has a wavelength associated with it that depends on the mass and the velocity of the particle. The faster the particle moves, the shorter its wavelength. Just as for photons, a smaller wavelength means more energy.

De Broglie's suggestion, like Einstein's theory of the photoelectric effect, was not a complete solution to the problem of electron motion. De Broglie gave no equation for his matter waves analogous to Maxwell's electromagnetic equations.

Three years later, Erwin Schrödinger came up with an equation to describe the motion of the matter waves. As reconstructed by physicist Leon Lederman, it happened this way:

> Leaving his wife at home, Schrödinger booked a villa in the Swiss Alps for two and a half weeks, taking with him his notebooks, two pearls, and an old Viennese girlfriend. Schrödinger's self-appointed mission was to save the patched up, creaky quantum theory of the time. The Viennese born physicist placed a pearl in each ear to screen out any distracting noises. Then he placed the girlfriend in bed for inspiration. Schrödinger had his work cut out for him. He had to create a new theory and keep the lady happy. Fortunately he was up to the task.[2]

Before we look at how quantum mechanics explained the line spectra of atoms, let's apply the matter wave idea and the Schrödinger equation to a simpler example, the harmonic oscillator. This special case appears often in physics, and it will be crucial to the discussion of relativistic quantum field theory.

Forget about quantum mechanics for a moment and picture an object sliding back and forth in a smooth bowl. It doesn't matter what the object is; let's say it's an ant on rollerblades. If you place the ant somewhere on the side of the bowl and release it, the ant will roll down to the bottom, then up the other side of the bowl. When it reaches the exact height at which you let go of it, it comes momentarily to rest, then rolls back down the other way until it returns to the point where you let it go. Assuming there is no friction, the ant will then repeat the same motion precisely, forever. You can start the ant from any height you like. If the shape of the bowl is just right, then the time it takes to go back and forth once will be the same no matter how high up the bowl you release the ant. You can picture how this works: from higher up, the ant has farther to go, but it will be going faster when it gets to the bottom. The effect of the longer distance is cancelled by the greater speed, resulting in the same travel time for any starting position. In this case, physicists call the system a *harmonic oscillator*.

As simple as it seems, this problem is one of the most important in all of physics. The reason is not that you commonly find ants rollerblading in specially shaped bowls, of course. First of all, it doesn't need to be an ant—any object will do. Then, almost any shape bowl will give you something very close to simple harmonic motion, if the oscillations are small—that is, if the object starts out near the bottom of the bowl. In fact, it doesn't have to be a bowl. Almost any situation where something oscillates will be described, at least approximately, by simple harmonic motion. A shaking leaf, a vibrating violin string, even the small variations of the moon's orbit due to the effects of distant planets, all can be described using simple harmonic motion. In the quantum world, too, we find it everywhere: the vibrations of atoms in a crystal, the states of the hydrogen atom itself, and the relativistic quantum field can all be expressed in terms of harmonic oscillators. Most importantly, the harmonic oscillator is a problem that can be solved! It may seem surprising, but most of the problems you run into in physics are unsolvable. The mathematics is simply too hard. Progress is made by finding approximate problems that keep the important characteristics of the original problem, but which are solvable. Much of the time, the solvable problem that you end up with is the harmonic oscillator in some guise.

The ant-on-rollerblades picture of the harmonic oscillator is a classical picture. That means we need to know the following:

- The initial location of the object—how far up the bowl we started the ant.

- The initial velocity of the object—in our case, the ant was at rest when we released it.

- The forces on the object—the force of gravity and the contact force between the bowl and the ant's rollerblades.

From this information, as we just discussed, we know what the ant is doing—forever. In the quantum world, the procedure is totally different. First of all, we need to represent the particle (ant) by a quantum field, or wave function. This field must be a solution of Schrödinger's equation. Just to show you what it looks like, here is the Schrödinger equation for the quantum field, denoted by ψ:

$$i\hbar \frac{\partial \psi}{\partial t} = -\frac{\hbar^2}{2m}\frac{\partial^2 \psi}{\partial x^2} + V\psi$$

You don't need to understand the mathematics of this equation, but there is one very important point about it to note: It introduces a new fundamental constant of nature, $\hbar$ (pronounced "aitch-bar"), known to physicists as Planck's constant. Like the speed of light, $\hbar$ can be measured by anyone, anywhere, at any time, and the result will always be the same. Whenever Planck's constant shows up, we know quantum mechanics is somehow involved.

The Schrödinger equation tells us how to take the particle wave and fit it correctly into the bowl. We find that the wave only fits at certain places in the bowl, giving a set of discrete energy levels.

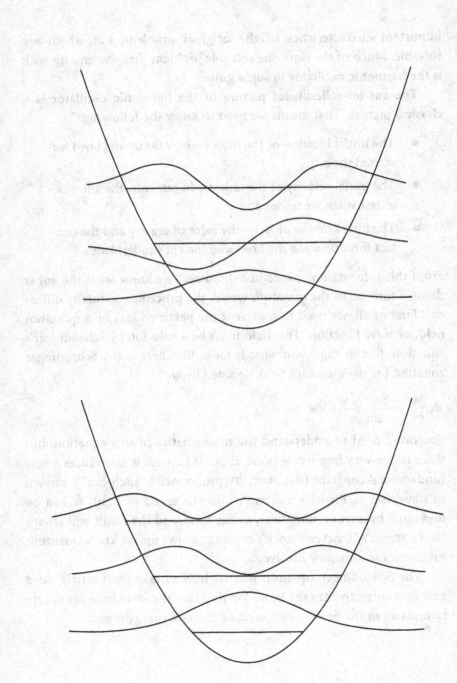

On the top, I plot the quantum field; on the bottom is the field value squared. According to the principles of quantum mechanics (specifically, principle (B) in the earlier list), the bottom graph gives the probability of finding the ant at a given location in the bowl. Look closely at the probability graph for the second energy level. Strangely, the probability of finding the ant exactly in the center of the bowl is zero! We thought we were describing something oscillating back and forth, but Schrödinger's equation tells us it can never be found in between! This is an unexpected consequence of the wave nature of particles.

Now consider the whole sequence of energy levels. The wave for the lowest level has just one bump, the next level has two, and so forth. As you try to cram more waves into the bowl, you have to go higher up the sides, that is, to higher energy. The Schrödinger equation tells us that only these energy levels are allowed. You can't have, for instance, one and a half waves across the bowl. There are no in-between energies allowed: The levels are discrete. Now, if we think about our rollerblading ant, discrete energy levels are quite strange. In classical physics, we can give the ant any energy we like; we simply start the ant at the appropriate height up the side of the bowl. Quantum mechanics seems to be saying we can't start the ant wherever we want on the bowl, only at certain discrete places. As odd as this may seem, it is exactly what we need to explain the sharp lines in atomic spectra.

The energy levels of the harmonic oscillator shown previously are perfectly evenly spaced. Suppose we use the harmonic oscillator to model the hydrogen atom: The "ant" is now an electron, and the "bowl" that keeps pushing the "ant" back towards the middle is the force of electrical attraction between the proton in the nucleus and the electron. If the electron is allowed only certain energies, it can't execute the death spiral we expected from classical electrodynamics. Instead, it must jump from one energy level to another. In each of these quantum jumps, the electron gives off energy in the form of light. When the electron reaches the lowest energy level (called the *ground state*), however, it can go no further. In classical mechanics, the electron can radiate away all its energy; in quantum mechanics, it ends up stuck in the ground state with nowhere to go.

What spectrum would we see from this "atom"? Because all the levels are the same distance apart (in energy), when the electron jumps from any level to the next lower level, it emits a photon of the same wavelength. This doesn't mean that there is only one line in the spectrum, however. An electron can also jump down two levels, or three levels, or more.

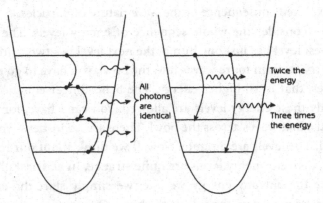

These jumps would generate photons with two (or three...) times as much energy as the photon generated by a jump of a single energy level. Separate the light from this "atom" by sending the light through a prism and you will see a series of evenly spaced lines, ranging from red to violet.

There is, however, no real atom that has this spectrum. For real atoms, the electric force between the nucleus and the electrons forms a "bowl" with a different shape, and therefore a different set of energy levels, than the harmonic oscillator.

For hydrogen, for instance, the visible lines are formed when an electron jumps down to energy level 2 from a higher energy level. Unlike the harmonic oscillator, the energy levels of hydrogen aren't evenly spaced. As a result, neither are the spectral lines.

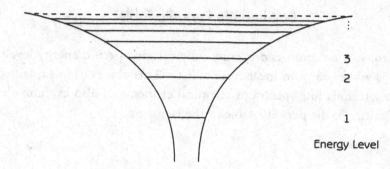

The energy levels are irregularly spaced, that is, there is more space between the first and second energy levels than there is between the second and third. As a result, electrons ending up in level 1 give off more energy than those ending up in level 2, and therefore emit photons with a shorter wavelength. These short wavelengths are called *ultraviolet light.* They are invisible to the eye, but can be recorded on photographic film. Electrons ending up in level 3 or any higher level, on the other hand, give off less energy, and thus emit longer wavelength photons. This part of the spectrum is called *infrared.* As with ultraviolet light, infrared light is invisible to the eye but can be recorded on film or with other specialized devices. The complete set of possible wavelengths forms the *line spectrum* of hydrogen.

Other chemical elements have different numbers of protons in the nucleus, causing different amounts of electric force, and so creating the spectral fingerprint characteristic of each element. The Schrödinger equation gives precise predictions for these energy levels, and therefore predicts the line spectrum that each element should produce. Atomic line spectra became the crucible in which quantum mechanics was tested and found brilliantly successful. For the most part, that is. Some small discrepancies remained, the so-called *fine structure* of the spectrum. These discrepancies would force physicists to look deeper, and would lead eventually to the even more accurate theory of quantum electrodynamics.

Building the Periodic Table

Electrons in an atom can occupy only certain specific energy levels: This is what quantum mechanics tells us. This observation explained the mysterious line spectra of chemical elements. It also explains the regularities in the periodic table of the elements.

Atoms, we know, are made of protons, neutrons, and electrons. The protons and neutrons in the atomic nucleus make up nearly all the mass of the atom. Electrons weigh only one-thousandth as much as a proton. The protons have a positive electric charge that is exactly equal in magnitude to the negative charge of the electron. Neutrons, as their name implies, are electrically neutral: They have no charge. The electrons are attracted to the positively charged nucleus, but they cannot fall all the way in because they are constrained to certain energy levels, as we have just seen. You need to know two other things about electrons. First, no two electrons can occupy the same quantum state. This property is called the Pauli exclusion principle, after Wolfgang Pauli. Particles that have this property are called *fermions*—we'll learn more about them in Chapter 8. The exclusion principle means, for example, that if you put two electrons in a box, they will never be found at the same place in the box. (This may seem obvious, but there are other types of particles that *can* be found at the same place.) Second, electrons have a property called spin. As its name implies, spin is connected with rotation. For an electron, however, spin can only take on two values, called *spin-up* and *spin-down*.

Now we're ready to start. We'll construct the periodic table from left to right across a row, and then proceed to the next row. This means we are going in order of increasing atomic number, which is simply the number of protons in the nucleus. Start with hydrogen (H): just one proton in the nucleus, and one electron (we are considering neutral atoms), which can go in any energy level it likes. Most atoms at room temperature are in their ground state, that is, their lowest possible energy level. Let's assume that's where hydrogen's electron is, in energy level 1. Next comes helium (He): two protons in the nucleus. We can throw in a handful of neutrons, too. In natural helium, two neutrons are the most popular configuration, but one neutron is also possible. These correspond to the isotopes helium-4 and helium-3; the number given counts the total number of protons plus neutrons. (Isotopes are chemically the same element; they only differ in the number of neutrons in the nucleus. The chemical properties of elements derive from the behavior of the electrons, which is unchanged by the addition of electrically neutral neutrons.) We add two electrons to make a neutral atom, and put both electrons into energy level 1, but one electron must be spin-up and one must be spin-down. This is because of the Pauli exclusion principle: Since both electrons are in the same energy level, they must have different spin directions in order to be in different quantum states.

After helium comes lithium (Li): three protons, typically four neutrons, and three electrons to make it neutral. The first two electrons can go into the lowest energy level (level 1), one up and one down, but the third electron can't go into level 1. Whether we assign spin-up or spin-down to the third electron, it will be in the same energy level and have the same spin state as one of the two electrons that are already in level 1. The Pauli exclusion principle requires that the third electron go into the next energy level, level 2. So, lithium has a full level 1 (two electrons) and a lone electron in the outer level 2. It is this lone electron that makes lithium chemically similar to hydrogen, which also has a lone electron. Lithium, therefore, goes in the next row of the table, underneath hydrogen. Beryllium (Be) comes next, with four electrons: two in level 1 and two in level 2. With boron (B), things get interesting: Schrödinger's equation allows for three states in energy level 2 in which the electron is rotating about the nucleus. Each of these states can hold two electrons (one spin-up and one spin-down, of course), so we get six elements to fill up the rest of the second row: boron (B), carbon (C), nitrogen (N), oxygen (O), fluorine (F), and neon (Ne). The next element, sodium (Na), has a lone electron in level 3, and so goes in the next row, under lithium. The third row fills in the same way as the second row, but in the fourth row we find an additional five rotational states, which gives room for the elements 21 (scandium [Sc]) through 30 (zinc [Zn]).

In 1871, the Russian chemist Dmitri Mendeleev organized the known elements into a periodic table. In order to place the elements with similar chemical properties in a single column of the table, Mendeleev had to leave gaps in the table. These gaps, he claimed, corresponded to elements that had not yet been discovered. For instance, no known element had the correct properties to fill the spaces in the table below aluminum (Al) and silicon (Si). He predicted the existence of two new elements, which he called eka-aluminum and eka-silicon. These new elements should resemble aluminum and silicon in their chemical behavior and both should be graced with atomic weights between the weight of zinc (Zn) and arsenic (As). Four years later, the element gallium (Ga) was discovered and found to have the properties Mendeleev had predicted for eka-aluminum. In 1886, germanium (Ge) was discovered, with the right properties to fill the space in the table below silicon. In hindsight, we can see these successful predictions as confirmation of the quantum-mechanical understanding of the atom. A similar process was to occur in the 1970s when physicists were classifying the plethora of particles being produced in particle accelerators.

In principle, Schrödinger's equation should be able to predict not just the structure and spectrum of atoms, but also the interactions between atoms. Chemistry then becomes part of physics. Instead of merely measuring reaction rates and energies between compounds, you could predict them by solving a quantum mechanics problem. In practice, this is extremely difficult to do, and many scientists are still attempting to do this today, as part of the field known as physical chemistry.

Chapter 4

(Im)probabilities

There are more things in heaven and Earth, Horatio,
Than are dreamt of in your philosophy.
—William Shakespeare, *Hamlet*

The Schrödinger equation gave quantum mechanics a solid mathematical foundation. But what did it mean to use a wave to describe a "solid" object like an electron? Physicists were used to classical physics, in which there is always a smooth transition from one state of motion to another. What were they to make of the bizarre quantum jumps from one energy level to the next? What was the relationship between the wave nature and the particle nature of objects? How were physicists to make sense of this absurd theory?

Half a Molecule Plus Half a Molecule Equals No Molecules

Successful explanation of the atomic line spectra provided good evidence for both the wave nature of electrons and the correctness of the Schrödinger equation. Could the wave nature of electrons be demonstrated more directly? An American experimenter, Clinton Davisson, decided to find out. Davisson realized that it would be impossible to perform a two-slit experiment for electrons like Young's two-slit experiment that demonstrated the wave nature of light. The wavelength of electrons, according to de Broglie's formula, was too small: You would need atom-sized slits to see electron interference. Davisson had discovered how to make crystals of metallic nickel, and he realized that an electron beam fired at such a crystal should behave in a similar manner as light behaves in the two-slit experiment. The nickel atoms in the crystal form an orderly array. When the electron beam strikes the surface of the crystal, each nickel atom reflects part of the beam. These reflected beams should combine like waves, according to de Broglie. That is, the reflected beam should display interference: At some locations the reflected waves should add (constructive interference), whereas at other locations they should cancel (destructive interference), just as in the two-slit experiment. Davisson successfully measured the interference pattern in 1927, providing the first direct experimental demonstration that particles have a wave nature.

Wave behavior of electrons may not seem so surprising. After all, no one has ever seen an individual electron; the electron is almost as much an "airy nothing" as the photon, the particle of light. Indeed, an electron is the smallest-mass particle of those that make up ordinary matter, and so is perhaps as close to nothing as a particle can get. However, similar experiments have since been performed with beams of neutrons, of atoms, even of whole molecules. Molecules are as solid as you can get: Everything we call solid is made of molecules, including your body and mine. If we can demonstrate wave behavior for molecules, the unavoidable conclusion is that everything in the universe must have a wave nature.

The largest molecules for which wave behavior has been demonstrated to date are called *buckminsterfullerenes*, or buckyballs, because of their shape, which resembles the geodesic domes built by architect Buckminster Fuller. Each molecule consists of 60 carbon atoms arranged in a soccer ball shape. Instead of a barrier with two slits, the experiment uses a grating having multiple slits 1/10 of a micron apart. These slits corral the buckyballs into a narrow beam. If the molecules behaved like peas shot from a peashooter, we would expect the detector to record molecules only in the direct beam path. Instead, the experimental results clearly showed the wave behavior of the molecules: spreading (diffraction) of the beam, and interference arising from the multiple slits of the grating.

We might be tempted to explain buckyball interference by assuming the molecules interact with *each other* in some way. Perhaps they bump into each other on the way to the detector, and that is what causes the peaks and valleys of the interference pattern. But this explanation doesn't work: the pattern is the same even if we send the buckyballs through one at a time. *Each* buckyball interferes with *itself*, as if it somehow passes through *all* the slits at once. The buckyball, though, is about 100 times smaller than the slit width, so saying that it "passes through all the slits at once" is like saying that a soccer ball goes through both goals simultaneously, at opposite ends of the field.

The interference experiment reveals the fundamental conceptual difficulty in quantum mechanics, the thing that induced physicists to use words like "crazy" and "absurd." Is an electron (or an atom, or a molecule) a particle or a wave? A single molecule can be isolated and weighed, it can collide with another molecule and bounce off, it can even be imaged using a scanning tunneling microscope. These properties make sense only if a molecule is a particle, something like a tiny billiard ball. On the other hand, the interference experiment only makes sense if the molecule can spread out like a wave in order to go through all the slits at once. Neither the wave model nor the particle model by itself can explain all of the experiments. We need a new conceptual model, something that is neither particle nor wave, something for which our everyday experience provides no analogy. That "something" is what I have been calling the quantum field. It is a field that obeys a wave equation (the Schrödinger equation) that gives us the properties of interference and so on, but it always comes in chunks— in any interaction, only a whole electron (or atom or molecule) is emitted or absorbed or detected.

If molecules show wave behavior, and everyday objects are made of molecules, why don't everyday objects exhibit that same behavior? Why can't we demonstrate interference effects using a peashooter and a board with two holes drilled in it? The answer lies in the size of Planck's constant, the new constant of nature in the Schrödinger equation. Together with the mass and the velocity of the particle, this constant determines the wavelength of the particle. Peas shot from a peashooter, for example, would have a wavelength of about 10^{-30} meter, much less than the width of a nucleus. To detect these interference fringes you would need to measure the position of the pea to this precision—an impossible task. In special relativity, it was the large speed of light that made special relativistic effects difficult to detect, and hence unfamiliar in our everyday experience. For quantum mechanics, it is the smallness of Planck's constant that removes wave phenomena so far from our everyday experience. Only at the atomic level do quantum effects become large enough to detect.

Getting Chancy

In the 1920s, physicists struggled to understand the quantum field and what it meant. Max Born, a German physicist, realized that everything made sense if the field was related to the *probability* of finding the particle at a given point in space.

Suppose we modify the two-slit experiment to trap the particle after it passes the barrier by placing a box behind each slit. Fire a single particle, an electron, say, toward the slits. The quantum field, we know, passes through *both* slits, and so part of the field ends up in one box and part ends up in the other box.

Born realized that the quantum field can't be the electron itself: An electron never splits up. You never find half an electron in each box. What you find instead is that, if you repeat the experiment many times, half of the time the electron ends up in one box and half of the time in the other box.

Probabilities are always positive numbers. If something is certain to happen (for instance, death or taxes), it has probability one, that is, 100 percent chance of happening. If something (say, your teenager being home by 10 P.M. on Friday night) is certain never to happen, it has probability zero. When there is less certainty, the probability is a number between zero and one: a fair coin has probability one-half of coming up heads. You combine probabilities by adding them. For instance, the probability that the coin will come up *either* heads *or* tails is equal to: (probability of heads) + (probability of tails) = 1/2 + 1/2 = 1. The coin must come up either heads or tails, so the total probability must be one.

Now, the probability can't be equal to the value of the quantum field, because probabilities are always positive numbers, whereas the quantum field can be positive or negative. Born discovered that the probability is equal to the square of the quantum field:

Probability = (Quantum Field)^2

This relationship makes interference phenomena possible. Suppose we tried to make a theory that used only the probability, instead of the quantum field value. Think again about the two-slit experiment. When only one slit is open, there is some probability that your detector will detect an electron. Now, open the other slit. According to the rules of probability, the probability of detecting an electron is now equal to the sum: (probability that electron comes from slit 1) + (probability that electron comes from slit 2). Since probabilities are always positive numbers, this sum is always bigger than the original probability. Using only probability, we can't explain what we found in the interference experiment: that opening the second slit can make the probability *decrease.*

Quantum mechanics gets around this by working with the quantum field instead of with the probability itself. We have to take the quantum field for an electron coming from slit 1 and add the quantum field for an electron coming from slit 2. Because the field values can be positive or negative, some places on the viewing screen will have a positive contribution from slit 1 and a negative contribution from slit 2, so that the sum is zero: These are the nodes (the valleys) of the interference pattern. The probability of detecting an electron at one of these nodal points is the total field value squared. Zero squared is equal to zero: We never see an electron at a node.

Adding quantum fields rather than probabilities is the fundamental feature of quantum mechanics that makes its predictions so counterintuitive. This principle is important enough to have a name: the *superposition principle*. The superposition principle says the way to find the probability of any outcome is to add all the quantum fields for all the possible routes to that outcome and then square the result. The superposition principle allows us to combine any two (or more) quantum states to obtain a new quantum state. For example, we can combine two energy levels of the harmonic oscillator to get a new state. The average energy of this new state lies between the energies of the original states. Measure the energy of the particle in the superposition state, though, and you will always observe one of the original energies, never the new, intermediate energy. The superposition state represents some probability of finding the particle in energy level 1 and some probability of finding the particle in energy level 2. We can adjust the probabilities at will, creating different superposition states with larger or smaller probability of either energy level, as long as the two probabilities sum to 1 (the particle must be in *some* energy level).

A superposition state is a very strange beast, for which there is no analogy in everyday experience. Suppose you wanted to paint your house red, but your spouse preferred blue. "No problem," says the painter, "I'll just mix the red with the blue." You agree reluctantly, fully expecting to come home to a purple house. Instead, you find that each time you return home the house is either red or blue. It's never purple. This isn't the way things work in our experience. We all know that if you mix red and blue paint you get purple paint, not paint that is sometimes red and sometimes blue. But in the microworld, where quantum mechanics rules, this *is* how things work. Using more red paint and less blue paint doesn't make the house redder, it's just red on more days. On blue days, the house is just as blue as if you hadn't used any red paint at all.

The superposition principle applies to any aspect of the particle, to any of its measurable properties. For instance, a particle's position can be in a superposition state. We already encountered one example: the modified two-slit experiment, in which the electron could end up in either the upper box or the lower box. When we look inside the boxes, we find the electron in one box or the other—not in some location in between. The velocity, the spin, even the charge of a particle can be in a superposition state. Superpositions of different charge states will be crucial for understanding quark physics.

The quantum field embodies the principles of quantum mechanics stated in the previous chapter:

A. The motion of any particle is described by a wave, which we call the quantum field.

B. The probability for the particle to be detected at a given point is equal to the square of the quantum field at that point.

C. The quantum field changes according to the Schrödinger equation.

Now, principles (A) and (C) are essentially the same as in classical physics. The Schrödinger equation, which tells how the quantum field changes in time, is quite similar to Maxwell's equations, which tell how the (classical) electric and magnetic fields change in time. There is no more randomness in this time evolution of the quantum field than there is in classical physics. The great departure from classical physics comes in the interpretation of the quantum field, given in (B). In classical physics, the fields were interpreted as real, physically existing entities spread throughout space. The quantum field, in contrast, is only an information wave. It doesn't tell us where the electron *is*. Rather, it summarizes everything we know about the electron. Quantum mechanics doesn't model the physical world, it reflects *what we can say* about the world.

The universe, it seems, does not admit of a complete description. The quantum mechanical description is the best we can do, and it gives only probabilities, not certainties.

I'm Not Sure If I'm Heisenberg

Quantum mechanics deals in probabilities. In some situations, though, the probability predicted by quantum mechanics is equal to one: There is no uncertainty about the outcome. For instance, an electron in the ground state of the hydrogen atom will be found to have the same energy each time the energy is measured. Could we avoid probabilities entirely by working only with states like this?

The Heisenberg uncertainty principle gives the answer: No! Physical quantities in quantum mechanics always come in complementary pairs: the more certainty we have about one quantity, the less certain we will be about some other quantity. The uncertainty principle is not an addition to quantum mechanics; it follows from the wave nature of particles. Suppose we try to trap a particle in a square well. Like the harmonic oscillator or the hydrogen atom, the square well has a set of discrete energy levels (calculable from the Schrödinger equation). In the figure that follows, the particle occupies a definite energy level, but its position is uncertain: it might be found anywhere in the well, or even a short distance outside the well. The particle's velocity is also uncertain: the possible values range roughly from zero to the value determined by the wavelength in this state:

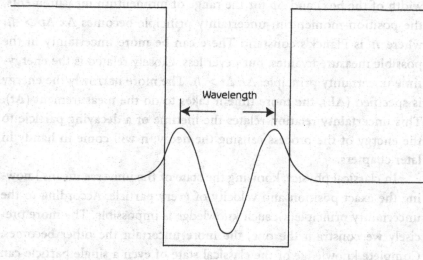

Let's reduce the uncertainty in position by decreasing the size of the box. Squeezing the box makes the wavelength decrease. Thanks to de Broglie, we know that a smaller wavelength corresponds to a larger velocity. Decreasing the uncertainty in position thus increases the range of possible velocities of the particle. We conclude that position and velocity are complementary physical quantities in quantum mechanics: decreasing the uncertainty of one quantity increases the uncertainty of the other.

The uncertainty relation just derived is the *position-momentum uncertainty principle*. Momentum being simply mass times velocity, a greater range of velocity implies a greater range of momentum. Writing Δx for the range of possible position measurements (that is, the width of the box) and Δp for the range of momentum measurements, the position-momentum uncertainty principle becomes $\Delta x \, \Delta p > \hbar$, where $\hbar$ is Planck's constant: There can be more uncertainty in the possible measured values, but never less. Closely related is the energy-time uncertainty principle: $\Delta E \, \Delta t > \hbar$. The more narrowly the energy is specified (ΔE), the more time it takes to do the measurement (Δt). This uncertainty relation relates the lifetime of a decaying particle to the energy of the process causing the decay; it will come in handy in later chapters.

In classical physics, knowing the state of the universe means knowing the exact position and velocity of every particle. According to the uncertainty principle(s), such knowledge is impossible. The more precisely we constrain the one, the more uncertain the other becomes. Complete knowledge of the classical state of even a single particle can never be obtained. The best we can do is to determine its quantum state, which only gives the probabilities of any of the measurable quantities.

The Dice-Playing God

For physicists from the time of Newton until the early twentieth century, the goal of physics had been the same—to predict the future. Consider your body: It is made of particles (atoms) interacting via fields (electromagnetic and gravitational). In principle, according to classical physics, a physicist could discover the complete state of your body (the position and velocity of every atom together with the field information), then use the equations of physics to calculate your every move—what time you will go to bed, when you will wake up, what your next words will be. This seems to make the universe a rather depressing place for a human—you are under the compulsion of the laws of physics. You have no choice about your next words, your next meal. Even if we cannot, in practice, calculate what you will do next, either because we don't yet know all the laws of physics, because we can't determine the exact location of every atom in your body, or because our computers can't handle the calculation, the very idea that such a calculation is possible in principle seems to leave the universe a very soulless place. Where is there any room for individual choice, for free will, in this picture? A human, it seems, is a mere robot following a complicated computer program.

Quantum mechanics completely undermined this mechanistic view of the universe, by removing not one but two of its foundations. First, according to the Heisenberg uncertainty principle, it is impossible, even in principle, to determine the exact position and velocity of each particle in your body. The best that can be done, even for a single particle, is to determine the quantum state of the particle, which necessarily leaves some uncertainty about its position or velocity. Second, the laws of physics are not deterministic but probabilistic: given the (quantum) state of your body, only the probabilities of different behaviors could be predicted. Quantum mechanics, it seems, releases us from robothood.

This radical change of perspective took a long time to be absorbed by the scientific community. The rules of the game had changed: The goal was no longer to predict the future completely, but to learn as much as can be known about the future, namely, the probabilities of some range of possible outcomes. Albert Einstein famously rejected the new view: "God does not play dice," he said. He claimed that quantum mechanics was incomplete; some deeper theory would be able to predict outcomes with certainty.

Think about rolling dice. If we knew the precise angle of throw, the exact effect of air resistance, and the friction with the table on which the die lands, then, in principle, we could predict with certainty which number will come up. In practice, the calculation is much too difficult, so the best we can do is a statement of the probability of each outcome: one in six (1/6) for each of the numbers on the die. Rolling dice is not *inherently* random; the outcome only seems random because of our ignorance of the little details, the hidden variables (like launch angle and friction) that determine the outcome of the roll.

Some laws of physics are probabilistic for the same reason as for the dice. Take, for example, the atomic model of gases. By averaging the motions of a huge number of particles, one can derive the ideal gas law and many other properties of the gas. Because the theory deals with the *average* properties, though, many questions we might ask about the individual molecules can only be answered in terms of probabilities. The atomic model can give us the probability that a particular gas molecule has a certain velocity, but we can't determine its actual velocity without knowing its entire history: its original velocity when it first entered the container, the angle and speed of every collision with the walls or the other molecules. If we knew those hidden variables, though, we could determine the exact velocity of the molecule. Einstein, and other physicists since him, wondered if quantum mechanics could be like rolling dice. Perhaps it only gives probabilities because we are ignorant of some of the hidden variables. Perhaps it is only an average theory, like the atomic model. Perhaps a deeper theory, one that included those variables, would give definite predictions.

Most physicists ignored Einstein's objections and accepted the arguments of Niels Bohr and others that quantum mechanics was complete. In the meantime, quantum mechanics racked up one successful prediction after another. Then, in 1967, John S. Bell published a paper that resolved the issue. Bell proved that any theory that involved hidden variables as envisioned by Einstein, and that excluded faster-than-light effects, *must* make predictions that conflict with the predictions of quantum mechanics. Quantum mechanics cannot simply be incomplete in the way that the atomic model is incomplete. If the hidden variable idea is right, then either quantum mechanics or special relativity is wrong. Bell's paper received little attention at first, but in 1969 it was realized that it was possible to directly test the conclusion experimentally. Since then, many further tests have been done, and quantum mechanics has won resoundingly. Bell's paper is now recognized as a fundamental advance in our understanding of quantum mechanics.

Let's delve a little more deeply into the meaning of Bell's discovery. Here's one of the basic equations of classical physics for the motion of an object with constant velocity: $x = vt$. In this equation, x represents the position of the object (measured from some reference point that we label $x = 0$), t represents the elapsed time, and v stands for the velocity of the object. For instance, a car traveling at 60 miles per hour will cover, in two hours, a distance $x = vt = (60 \text{ miles/hour}) (2 \text{ hours}) = 120$ miles. This may seem very straightforward and simple, but we need to recognize that something very subtle has taken place. We have taken an equation ($x = vt$), an abstract mathematical expression, and related it to a physical phenomenon (motion of an object), by way of an interpretational scheme ($x \leftrightarrow$ position of object, $t \leftrightarrow$ elapsed time, $v \leftrightarrow$ velocity of object). This was the great triumph of physics from Galileo in the sixteenth century through Maxwell in the nineteenth: that physical phenomena could be transformed, via some interpretational scheme, into mathematical relationships.

In cases like the simple equation $x = vt$, the interpretation is straightforward. We have an intuitive notion of the concepts *position of the object* and *time elapsed*, and we can make those ideas more precise if necessary. For example, *position of the car* could be specified as the *distance from the tip of mile marker 1 to the license plate on the rear bumper, measured in miles*. But even in classical physics, the interpretation scheme can be quite complicated and removed from everyday experience. The electric field of Maxwell's equations, for example, is not something that can be experienced directly. We need an interpretational scheme that goes something like this: The electric field at a location represents the force that would be felt by an object carrying one unit of charge if placed at that location, and if the introduction of that charge didn't affect the distribution of the other charges that created the field. The interpretation here involves a hypothetical situation, the introduction of an additional electric charge, which may be difficult if not impossible to accomplish in practice. If, for example, we are concerned with the electric field due to a charged conducting sphere, we know that introducing an additional charge in the vicinity of the sphere will cause the charges on the sphere to move around. There is no practical way to "nail down" the charges as required by the interpretational scheme. Still, this interpretation of the electric field is extremely useful; in fact, it works so well that physicists have no hesitation in thinking of the electric field as a real, physical entity, ontologically on a par with atoms, molecules, dust specks, and cars. The electric field (more precisely, a change in the electric field) moves with a measurable speed: the speed of light. The field also carries energy; for instance, the electromagnetic (light) energy reflected from this page carries energy to your eyes, which convert the light energy into nerve impulses that your brain can decode. Suppose we wanted to avoid the electric field concept entirely, and use equations that refer only to positions and velocities of "real" objects: stars, planets, rocks, and particles. Then we would find that energy disappears from the light bulb illuminating this book, reappears briefly on the page, then disappears again before reappearing in the rods and cones of your eyeball. Now, there is nothing logically wrong with an interpretation that avoids the electric field concept entirely, in which energy leaps into and out of existence like this. However, the electric field concept is so useful, and the conservation of energy so compelling theoretically and so well established experimentally that it seems eminently reasonable to consider both energy and electric fields as real in the same sense that atoms (and cars) are real.

What about quantum mechanics? Is the quantum field real? True, it describes the motion of matter (the electron, say) and therefore energy (the electron's kinetic energy, for example) from one place to another and so, like the electric field, would seem to be real enough. But the probability interpretation makes this view difficult to maintain.

Consider again the modified two-slit experiment discussed earlier. The incoming quantum field splits into two parts; there is an equal probability for the electron to end up in either of the two boxes that we placed behind the screen. Now suppose we insert a detector behind one of the slits. As the quantum field passes the detector, the detector either clicks, registering that a particle was in that box, or it doesn't click, in which case the particle must have been in the other box. There is no longer a 50-50 probability. The probability is now 100 percent for one of the two boxes and 0 percent for the other box. This is sometimes called the collapse of the wave function: Part of the quantum field disappears and the quantum field elsewhere instantaneously changes. Now imagine extending the two boxes in different directions. You can make them as long as you want, so the two arms of the apparatus could be separated by miles, or light-years, when the collapse takes place. If the quantum field is a real physical object, we seem to have a violation of special relativity—a physical effect that travels faster than the speed of light.

The violation is only apparent, however. Many years of careful study of such situations have revealed that this kind of instantaneous rearrangement of the quantum field cannot be used to send a signal faster than the speed of light. It is only through our interpretation of the quantum field as a physical object that a faster-than-light effect seems to arise. If we want to avoid faster-than-light effects (illusory though they may be), we are forced to declare that the quantum field is not a real, physical object. But then, what is it? Recall that the field tells us the probability of finding the electron at a particular position. If the quantum field is not a physical entity, perhaps it is a mathematical device that encodes all our knowledge about the electron. It is an information wave. In this view, the particle (the electron) is the independently existing physical entity, and the quantum field summarizes everything we know about the particle—the combined effect of everything that has occurred in the particle's past. Then, the collapse of the quantum field involves no physical effect that moves faster than the speed of light; it is merely a reshuffling of information that occurs whenever new information is obtained, as when Monty Hall opens Door Number Three to show that the grand prize is not there, and the probabilities of the prize being behind Doors One and Two immediately change.

The main question in this information interpretation of the quantum field is this: How does the electron know where it should be? If the quantum field is not a physical object, why is the electron's probability always governed by the quantum field? "It just is" is a possible, but not very satisfactory, answer. We would rather have a theory that tells us how the real, physical objects behave than one that is a mere calculation tool for probable outcomes.

Let's suppose, then, that there is a deeper theory of the electron. Perhaps the electron has some unknown properties that determine its motion, or perhaps it interacts with a guiding wave that tells it where to go. This is where Bell's result comes in. We can say with complete confidence that, no matter what unknown properties or guiding waves we introduce, the new theory will either involve physical effects that move faster than the speed of light, or else it will be in conflict with the results of quantum mechanics. We see there are three possible options:

1. Quantum mechanics may be violated. Since Bell's paper was published, many experiments have searched for such violations. Some early experiments seemed to show that, indeed, quantum mechanics was flawed, but all of the more recent and more accurate experiments have resoundingly supported quantum mechanics.

2. Special relativity may be violated, but in such a way that no signals can be sent faster than light speed. A small group of physicists is trying to develop theories along these lines. Unless quantum mechanics is proven wrong by experiments (see option 1), these theories must agree in the end with quantum mechanics, so the attitude of the majority of physicist to these attempts is "Why bother?"

3. We can avoid questions of interpretation by throwing up our hands and saying "It just is" the way quantum mechanics says it is. This is the approach taken by most working physicists today.

This is the rather sorry state of our understanding of quantum mechanics. As the theory (or rather, its extension to relativistic quantum field theory) continues to pass one experimental test after another, we still lack an agreed-upon interpretational scheme for the theory. Think about our earlier example using the equation $x = vt$. The interpretational scheme involved here is this:

Mathematical Variable		Physical Quantity
x	$\leftrightarrow$	Position of object
v	$\leftrightarrow$	Speed of object
t	$\leftrightarrow$	Elapsed time

What about quantum mechanics? The equation we use is the Schrödinger equation:

$$i\hbar \frac{\partial \psi}{\partial t} = -\frac{\hbar^2}{2m}\frac{\partial^2 \psi}{\partial x^2} + V\psi$$

Here, the interpretational scheme is as follows:

Mathematical Variable		Physical Quantity
x	$\leftrightarrow$	Position of electron
t	$\leftrightarrow$	Elapsed time
m	$\leftrightarrow$	Mass of electron
V	$\leftrightarrow$	Potential energy of electron
ψ	$\leftrightarrow$	???

For the first time in physics, we have an equation that allows us to describe the behavior of objects in the universe with astounding accuracy, but for which one of the mathematical objects of the theory, the quantum field ψ, apparently does not correspond to any known physical quantity.

The implications are deep and far-reaching, not just for physicists and their practice, but also for how we understand the world around us. The old view of a clockwork universe is comforting in its comprehensibility. It is easy for a physicist to imagine taking apart the universe like a clockmaker taking apart a clock, pulling electrons out of atoms like pulling a gear from among its neighbors. Each piece fits together with those around it, and by understanding the pieces and how they interact, the physicist can understand the whole. On the other hand, this is a cold, mechanical view of the world: every event, even every human action and emotion, is determined by the interactions of the constituent pieces, and those interactions are fixed by the preceding interactions, and so on, back to the beginning of the universe.

The quantum mechanical universe is very different from the clockwork universe, although the manner in which it differs depends on which of the three previous options is chosen. I will ignore the first possibility, that quantum mechanics is simply wrong. Quantum mechanics, and its descendant, relativistic quantum field theory, which is built on the framework of quantum mechanics, are simply too well supported by experiment. Even if the specifics of the Standard Model prove incorrect, the framework of quantum mechanics will certainly stand for many years to come.

The second option, of instantaneous (or at least faster-than-light) communication between particles, leads in the direction of a mystical, New Age view of the universe that makes some physicists uneasy. The argument goes like this: Because of the strange, non-local nature of the quantum field that we saw earlier, any two electrons that interact carry a strange sort of correlation, an instantaneous connection that can be ascribed neither to a property that the electron has of itself nor to a communication (in the usual sense) between the electrons. But before these two electrons interacted, they interacted with other electrons, and before that with other electrons. It seems unavoidable that all the electrons in the universe are caught in this web of interactions, so that any given electron (or indeed any other particle) has a kind of mystical, instantaneous connection with every other object in the universe. This is the view of quantum mechanics that led to comparisons with Eastern religions and was popularized in the 1970s and 1980s in books like Fritjof Capra's *The Tao of Physics*. This view, or at least the connection with religious mysticism, is usually derided by physics writers today, but it remains true that in some approaches to quantum mechanics, these long-distance connections are real and physically important.

The third option denies physical reality to the long-distance correlations, in the sense that there is no separate, physical field or other entity postulated to account for the correlations. However, by doing so, it leaves the correlations completely unexplained. Quantum mechanics can tell us what correlations to expect (and, lo and behold, the predicted correlations are found experimentally), but it leaves us utterly without a mechanism, without an answer to the question "Why?" It is not that this view denies the existence of an independent physical reality; it is just that nothing in our theoretical framework can be identified with that reality. The quantum mechanical formalism tells how to compute the probabilities for the outcome of any experiment, but it is only a calculational tool that magically produces the right answer; the mathematics is not a direct reflection of the underlying physical reality as it was for Newton and Maxwell.

Philosophically, the most astonishing thing about quantum mechanics is the extent to which it protects us from the existential despair of the clockwork universe. Nor is it necessary to speculate on the possibility of long-distance correlations providing a route for mystical spiritual interconnectedness, or other metaphysical mumbo-jumbo. In order to speak of a "quantum state" at all, we need to have a repeatable sequence of physical processes that produce the state. This is how a quantum state is defined. It is simply impossible to apply the quantum state concept to the unique configuration of electrons, protons, neutrons, and photons that is a human being. No repeatable process will produce another copy of me at the moment of writing this sentence, or of you at the moment of reading it. Furthermore, even if such a description were possible, quantum mechanics gives us only the *probabilities* of various outcomes. Given the quantum state of the reader at, say, noon today, we could (in principle—such a thing is beyond the capabilities of modern computers the way the Andromeda galaxy is beyond your neighbor's house) compute a 6 percent probability that you will have spaghetti for dinner tomorrow, 3 percent probability of macaroni and cheese, 5 percent for hamburger.... Such predictions are no assault on free will; similar predictions could be made simply by looking at your past eating habits. It is almost as though the rules of the universe were designed to protect our free will.

Physicists are split on which of the three views of quantum mechanics they believe. Many, perhaps most, prefer not to worry about the problem as long as they can calculate the answers they need. A few brave souls struggle with the questions of interpretation or construct competing models that act as foils against which quantum mechanics can be tested. It is clear, though, that quantum mechanics replaces the clockwork universe, in which the future can be determined to any desired precision, with a world of unpredictability. Not just complex systems like the weather or the brain, but the simplest imaginable systems, a single particle, for instance, are subject to this randomness.

Quantum mechanics was not the end of the story, however: problems both theoretical and experimental remained. On the theoretical side, quantum mechanics takes a Newtonian view of space and time—it is incompatible with special relativity. On the experimental side, the details of the atomic line spectra turned up surprises. Some lines split into two or more lines when new techniques allowed a closer inspection. Other lines were shifted slightly from their expected positions. Small though these differences were, they would spark a new revolution. The grand synthesis that finally harmonized special relativity and quantum mechanics would also solve the experimental puzzles and lay the foundation for the Standard Model. That synthesis is called relativistic quantum field theory.

Chapter 5

The Bizarre Reality of QED

> The wicked regard time as discontinuous, the wicked dull
> their sense of causality. The good feel being as a total dense
> mesh of tiny interconnections.
> —Iris Murdoch, *The Black Prince*

Imagine that you are getting ready to go to work one morning, and you suddenly realize you need to drop off the kids at school, mail a letter, and pick up money from the cash machine. You're already running late, so you do the only possible thing: split yourself into three people. One of you drops the kids off, another swings by the post office, and the third goes to the bank and gets the cash out. Finally, you all arrive at work, reunite, and go about your business as a whole person.

This may sound like a bizarre science fiction fantasy, or maybe a drug-induced nightmare hallucination, but according to our best understanding of subatomic physics, this is what the world of elementary particles is like. In fact, it's even worse (or better!) than that: Different versions of you would make stops in Aruba, the Alps, the moon, and Woonsocket, RI. It may seem like you would need to travel faster than the speed of light to get to the moon and still make it to work on time, thus violating special relativity. In fact, it is the confluence of special relativity, quantum mechanics, and field theory that leads us to this bizarre picture of how the world works.

In the 1930s and 1940s, physicists came to accept special relativity as the correct description of space and time. At the same time, quantum mechanics was giving predictions about atomic spectra that were in excellent agreement with experiments. There were two problems, though, one experimental and one theoretical. The experimental problem was that, as better and better techniques were developed for measuring the lines of atomic spectra, details of the spectra were found that weren't explained by quantum mechanics. The theoretical problem was that quantum mechanics was a nonrelativistic theory—in other words, the equations were in conflict with special relativity. The resolution that was painstakingly worked out over the next 20 years would weave together the three themes of field theory, special relativity, and quantum mechanics into a harmonious whole, a structure that would be the basis of the most successful scientific theory of all time.

The Mystery of the Electron

P. A. M. Dirac took the first step toward making quantum mechanics compatible with special relativity in 1928. The Dirac equation replaced the Schrödinger equation of quantum mechanics. It was perfectly consistent with special relativity, and almost as a bonus, it explained the electron's spin. Spin, you will recall from Chapter 3, was the mysterious property of electrons that let us put two electrons in each energy level, rather than one (as the Pauli exclusion principle would seem to demand).

All electrons are perfectly identical; there is no individuality among them. They all have precisely the same mass, the same electric charge, and the same spin. You never see an electron with half the mass, or a slightly larger or smaller charge. These properties *define* an electron: If you see a particle with the right mass, charge, and spin, it is an electron. If any one of these quantities is different, it's not. The same statement, modified to include other types of charge, can be made about any other elementary particle.

What is this thing called spin? The electron's spin, as the name implies, has to do with rotation. Take a beam of electrons that are all spinning in the same direction and fire it at, say, a brick. If you could keep this up for long enough, and if there were no other forces acting on the brick, the electrons would transfer their rotation to the brick, and it would begin to rotate. So a spinning electron behaves in some respects like any other rotating object. In other respects, though, the electron's spin seems a bit odd. For instance, the spin rate never changes—it never speeds up or slows down, just as a spinning top would neither speed up nor slow down if it were on a perfectly smooth, frictionless surface. Only the direction of rotation, called the spin axis, can change.

It is tempting to think of electrons as tiny spinning balls of charge. There are problems with this picture, however. Think of how an ice skater speeds up as she pulls her arms in. The smaller she makes herself, the faster she spins. Now, no one has measured the size of an electron, but there are experiments that give it an upper limit. According to these experiments, the electron is so small that, to have the known value of spin, the surface of the "ball" would have to be moving faster than the speed of light. This, of course, is impossible. We are forced to conclude that an electron is not a tiny, spinning ball of charge. Nor can the model be fixed by replacing the ball with some other shape. Whether ball shaped, donut shaped, or piano shaped, the picture of an electron built out of a collection of even smaller charges simply doesn't work.

Well, then, what is an electron? What has charge and spin but isn't a spinning charge? The only option is to picture the electron as a truly fundamental particle, a pure geometric point having no size and no shape. We simply have to give up the idea that we can model an electron's structure at all. How can something with no size have mass? How can something with no structure have spin? At the moment, these questions have no answers. There is nothing to do but accept that the electron does have these properties. They are merely part of what an electron *is*. Until we find some experimental evidence of electron structure, there's nothing else to say about an electron.

Since all electrons have exactly the same charge, and all other free particles that have been detected have whole number multiples of that fundamental charge, we can simply say the electron has charge -1. Protons, for instance, have charge $+1$—the same value of charge as the electron, but positive instead of negative. (Physicists believe that protons and neutrons are made up of quarks having 1/3 or 2/3 of the electron's charge. This of course makes the smaller value the truly fundamental unit, but instead of rewriting all the physics books to say the electron has charge -3, physicists have stuck with the old definition.) The electron's spin, in fundamental units, is 1/2. There is no known fundamental mass, however; the electron's mass is an inconvenient 9.11×10^{-31} kilograms. This is as small compared to your body mass as an amoeba is compared to the entire earth.

It is impossible to measure the electron's spin directly, as suggested earlier, by shooting a beam of electrons into a brick and measuring how much rotation the brick picks up. The electron's spin is much too small compared to the inertia of any such object for this method to work. The spin was deduced indirectly from careful measurements of the light spectrum produced by atoms. The rules of quantum mechanics, as you will recall, require that the electrons around an atom occupy distinct energy levels. Now, an electron has both charge and spin, and according to Maxwell's equations it should behave as a tiny magnet. If the atom emitting the light is immersed in a magnetic field, the energy levels are shifted by the interaction between the electron's magnetic properties and the magnetic field. The energy level shift shows up in the spectral lines; this phenomenon is known as the *Zeeman effect*.

By immersing atoms in magnetic fields and through other experiments, physicists deduced that the electron's spin can point in one of only two directions: either in the same direction as the magnetic field (called spin-up) or opposite to it (spin-down). It may seem strange that the spin can't point in any arbitrary direction; after all, there are no restrictions on how we orient the spin axis of a top or a gyroscope.

The quantization of spin, like the quantization of the energy levels, is a consequence of quantum mechanics, however. Just as the energy levels of a (quantum) ant rollerblading in a bowl could only take on certain discrete values, the spin direction of the electron has only two possible values. The explanation of the Zeeman effect in terms of quantum principles was one of the early successes that convinced physicists that quantum mechanics was correct.

The Dirac equation was brilliantly successful: It agreed with special relativity, it got the electron's spin correct, and it explained results like the Zeeman effect that had seemed inexplicable with the old, nonrelativistic Schrödinger equation. The new equation brought some puzzles of its own, however. The first problem Dirac noticed was that, in addition to the solutions of the equation that described electrons so well, there were other solutions that had negative energy. In fact, there were infinitely many solutions with negative energy. This was unacceptable. To see why, remember what a quantum system is supposed to look like—the previously discussed harmonic oscillator, for instance. An electron in a higher-energy state can drop to a lower-energy state by radiating energy away. But, once in the lowest-energy state, called the ground state, there is nowhere to go. The electron must sit there until it gets zapped with enough energy to move up again. In contrast, Dirac's equation gave a picture of a bottomless pit, where electrons can keep radiating and dropping, radiating and dropping, without any end. In this way, a single electron could produce an infinite amount of energy output, without any energy input. This was in clear contradiction to experimental fact—something was wrong.

Dirac soon found a way around this problem. In hindsight, we can say that his solution was only a preliminary one, a first step toward relativistic quantum field theory. When we talked about the periodic table, we used Pauli's exclusion principle, which said that there can be only one electron in any quantum state—if the state is already filled, no more electrons can come in. Dirac made a bold assumption—that the infinitely many negative energy states were all filled, everywhere in the universe! This meant that there was an infinite amount of mass and infinite (negative) electrical charge at each point in space. According to Dirac, though, this infinite sea of mass and charge is normal and so we don't notice it. This may seem to be a lot to sweep under the rug, but it seemed to work. However, there is a situation where we would notice the sea: when one of the particles in the negative sea gets zapped with enough energy to lift it to a positive energy state. Picture a can packed tightly to the brim with marbles. If you remove one marble it leaves a hole behind. Similarly, removing an electron from the sea leaves an electron-shaped hole in the sea. Now, in the normal condition, all the negative energy states are filled. So, a hole in the sea, which is a missing negative charge, must look to us like a positive charge. As a result, if you dump enough energy at any point in space, you produce two objects *ex nihilo*: an electron and a positively charged hole. What's more, nearby electrons in the sea can move over to fill the hole, but that will just move the hole elsewhere. The hole can move around. Dirac found that the hole behaves exactly like a particle—a positively charged particle.

This was tremendously exciting for Dirac. At this time (1931), there were only two known subatomic particles: the electron, with charge −1, and the proton, with charge +1. Dirac's equation seemed to be describing two such particles. Maybe the holes were actually protons. If so, the Dirac equation would be a complete unified theory of elementary particles. Unfortunately, this solution didn't work. Dirac soon realized that the holes had to have the same mass as the electrons. Protons, at 2000 times the electron's mass, were just not going to work. (The discovery of the neutron a year later would have changed the picture anyway.) Dirac had no choice but to predict the existence of a particle that no one had ever seen or detected, a particle with the same mass and spin as an electron, but with positive charge. This radical step was found to be justified in 1932. Dirac's positive electrons, dubbed *positrons*, were detected in the tracks of cosmic rays in cloud chambers.

Cloud chambers had been used for some years as particle detectors. The cloud was created by spraying a fine mist of water into the chamber. When particles passed through the cloud, small droplets of water condensed around their path. One could then take a photograph and make measurements of the paths. A magnetic field bends the particles' paths (as we know from Chapter 1), so by surrounding the chamber with a powerful magnet, even more information about the particles could be gleaned.

By using radioactive substances as sources, physicists had learned to identify the cloud chamber tracks of the known particles. Even without a radioactive source around, though, they saw tracks. These tracks apparently rained down from the sky, and so were named cosmic rays. Carl D. Anderson, a young physicist working at Caltech, noted that in some of his cloud chamber photographs there were "electrons" that seemed to curve the wrong way in the magnetic field. Perhaps, though, they were normal electrons traveling the other way—coming up from the ground instead of down from the sky. To resolve the question, Anderson inserted a metal plate into the chamber. Particles passing through the plate would be slowed down, and slower moving particles would be curved more sharply by the magnetic field. With the plate in place, all Anderson had to do was wait until one of the backward-curving particles passed through it. He finally obtained a photograph of such an event that settled the matter. Astonishingly, the particle had the mass of an electron but a positive charge. Anderson had discovered the positron, predicted by Dirac two years earlier.

Anderson's experiment confirmed the existence of what we now call *antimatter*. The positron is the antimatter partner of the electron. Every particle discovered so far has its antimatter partner, called the *antiparticle*. Anderson's discovery also confirmed Dirac's place in the pantheon of physics. For the first time (but, as we will see, by no means the last), a particle predicted from purely mathematical considerations had been found to exist in reality. In a few years' time, both Dirac and Anderson were awarded Nobel prizes for their achievements.

Suppose we have an electron and a hole. The electron can fall into the hole, but when it does so it must radiate away some energy. We are left with no particles, just the leftover energy. The electron and the positron have annihilated each other; the mass of both particles has been entirely converted into energy. Since each particle has rest energy equal to mc^2, particle-antiparticle annihilation always produces at least twice that much—an energy of $2mc^2$. More energy can be produced if either particle is in motion before the collision, because there is then some kinetic energy in addition to the rest energy of the two particles.

As the Dirac equation was successfully applied to describe more and more of the particles that were being discovered, it was realized that not just electrons, but all particles must have an unidentical twin: a particle with the opposite charge, but otherwise with the same properties as the original particle. These antiparticles could annihilate with normal particles, leaving only energy behind. Because the antiparticles behave in the same way as regular matter, they can, in principle, combine to form antiatoms and antimolecules. Antichemistry would be identical to regular chemistry, and so the same sorts of objects and creatures that we have in our matter world could, in principle, exist in a world made entirely of antimatter.

Imagine an anti-earth somewhere, with antipeople on it made out of antiprotons, antineutrons, and antielectrons (that is, positrons). On this planet, there may be an antiyou who becomes an astronaut and comes to visit you on earth (perhaps using a matter drive?). On landing, however, the spaceship immediately explodes in a tremendous matter-antimatter annihilation. The energy released would be equal to several thousand nuclear weapons.

Once we realize that there is a symmetry between particles and antiparticles, it makes more sense to treat them symmetrically, rather than thinking in terms of Dirac's hole picture. That is, instead of thinking of electrons as fundamental and positrons as missing negative-energy electrons, we think of electrons and positrons as equally fundamental. The rest of this chapter will show what picture emerges when we do this.

Our world is full of protons. Where are the antiprotons? Of course, we do not expect to see any antiprotons here on earth—they would quickly run into a proton and annihilate. Is there an antimatter planet somewhere in the universe, perhaps orbiting an antimatter star in an antimatter galaxy? Are there regions of the universe that are mostly antimatter, just as our neighborhood is mostly matter? If there were, then wherever a matter region borders an antimatter region there should be a mixing region where the particles and antiparticles meet and annihilate releasing energy (in an amount equal to $2mc^2$ for each annihilation). Astronomers would see this radiation as a bright spectral line with known energy, so it would be very easy to detect. In spite of many hours spent in telescopic observations, though, no evidence of an antimatter region of the universe has ever been found. We must conclude that all the stars we see, all the distant galaxies, are made from normal matter, not from antimatter.

This is puzzling: If there is a complete symmetry between particles and antiparticles, why aren't there as many antiparticles in the universe as particles? The lack of antimatter is a deep mystery that cannot be explained using the Standard Model. It implies that the particle-antiparticle symmetry is not quite complete. As we will discover, the very existence of large clumps of matter, such as galaxies and galaxy clusters, provides a hint of physics beyond the Standard Model.

Dirac Rules!

One day in the early 1940s, John Archibald Wheeler, then a young assistant professor at Princeton, called up the person he usually bounced his crazy ideas off of: Richard Feynman, at that time still a graduate student at Princeton.

Wheeler said, "Feynman, I know why all the electrons have the same charge and the same mass."

"Why?" Feynman asked.

"Because they are all the same electron!" replied Wheeler.[1]

What Wheeler had realized was that, from a mathematical point of view, a positron is the same as an electron traveling backward in time. We can draw it like this:

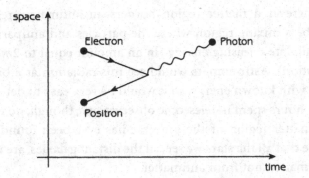

The arrow shows the direction the electron is traveling, so the line with the arrow pointing backward in time looks to us like a positron traveling forward in time. So, we see an electron and a positron that get closer and closer (in space) until they annihilate, and we are left with only a photon. Alternatively, we can think of the same picture as depicting an electron traveling forward in time, which then emits a spontaneous burst of energy and starts going backward in time.

To see how this explains why all electrons look exactly alike, look at this diagram:

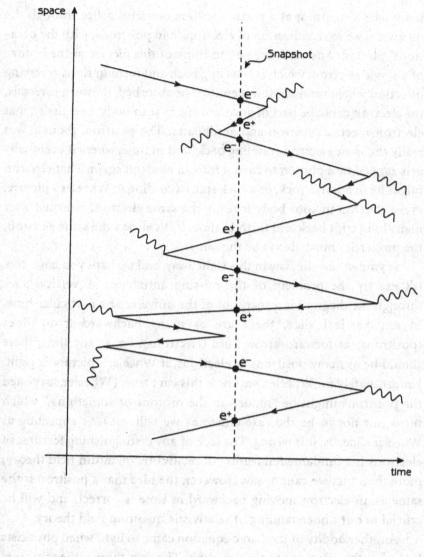

If we take a snapshot at a particular time (a vertical slice through the diagram), we see a collection of electrons and positrons, with the occasional photon. Alternatively, we can think of this picture as the history of a single electron, which is traveling back and forth in time, reversing direction whenever a photon is emitted or absorbed. Between reversals, the electron could be part of an atom, say, in your body. Eventually, that electron meets a positron and annihilates. The positron, though, was really the same electron traveling backward in time, where it eventually gets zapped by a photon to turn it into an electron again. That electron might be in a star, a rock, or a rock star. According to Wheeler's picture, every electron in your body is really the same electron, returned after many long trips back and forth in time. If it's always the same electron, the properties must always be the same.

Feynman saw the flaw in this right away, and perhaps you have, too. It's exactly the problem of the missing antimatter. A vertical slice through the diagram is a snapshot of the universe at a particular time. In any vertical slice, there are as many backward-arrow lines (positrons) as forward-arrow lines (electrons). So, at any time, there should be as many positrons as electrons, if Wheeler's picture is right. Unfortunately for Wheeler, we know this isn't true. (Wheeler suggested the positrons might be "hidden in the protons or something," which turns out not to be the case either, as we will see.) As appealing as Wheeler's idea is, it is wrong. The lack of any distinguishing features of electrons is a fundamental feature of relativistic quantum field theory; more than that we cannot say. However, the idea that a positron is the same as an electron moving backward in time is correct, and will be crucial to our understanding of relativistic quantum field theory.

Another oddity of the Dirac equation came to light when physicists looked at the solutions to the equation. The first thing a theorist does when confronted with a new equation is to solve it for the simplest cases imaginable. One of the simplest possibilities is to imagine firing a beam of electrons at a barrier. Since electrons are charged particles, we can suppose the barrier is an electric field that is strong enough to repel the electrons. The solutions showed that there were more electrons in the reflected beam than in the incoming beam. This is like saying you were batting tennis balls against a brick wall and for every tennis ball you hit, two came back at you. The resolution of this conundrum comes when we look "inside the wall"—behind the barrier. For each extra electron in the reflected beam, there is a positron that travels off in the other direction, through the wall.

Here's what's actually happening: The barrier is creating pairs of electrons and positrons. The electrons come back at us, like the extra tennis balls. Because of their positive charge, the positrons react differently to the electric field that forms the barrier. The same field that repels the electrons attracts the positrons, and they move off through the wall.

It probably seems like we're getting something for nothing: For each electron going in, two electrons and a positron come out. The source of the extra particles turns out to be the wall itself, or rather, the electric field that creates the wall. Using Dirac's hole picture, we can say that energy from the electric field pops an electron out of the infinite sea of negative-energy electrons, leaving a hole behind. The remaining electric field propels the electron in one direction (becoming the extra tennis ball) and the positron in the other.

This is an example of what physicists call a *thought experiment*. It is an idealized situation that can't easily be turned into a real experiment. The difficulty in this case comes in constructing the wall off which the electrons bounce. A wall built of, say, lead bricks will not be very effective. To an electron, the atoms of the brick look like mostly empty space. The electron won't bounce off the lead brick; it will penetrate it, scattering off other electrons in the brick until its energy is dissipated. Rather, the wall must be created using an electric field, because that is the only thing we know of that repels electrons. But where to get the electric field? Electric fields come from an accumulation of electric charges, that is, electrons! In other words, the only thing that an electron will bounce off is another electron. The real experiment that is closest to our thought experiment, then, is one in which an electron collides with another electron.

Well, what happens when electrons collide? Do we actually see more electrons flying out than we put in? Yes! The phenomenon is known as *pair production*: electron-positron pairs spontaneously leaping into existence. In fact, this is the main method physicists use to produce positrons: fire a beam of electrons of sufficiently high energy at some target and collect the positrons that fly out. It should come as no surprise that "sufficiently high energy" translates to "more than $2mc^2$"—twice the rest energy of an electron.

The Dirac equation didn't provide a complete theory of pair production; it was invented as an equation for the behavior of a single electron. Because of pair production, a one-particle theory is insufficient. Only a theory capable of dealing with electron-electron interactions, and with interactions of electrons with the photons of the electromagnetic field, would be capable of describing the real world.

Let's pause and think about what we've accomplished. Dirac's equation has brought two of our themes, quantum mechanics and special relativity, into an uneasy harmony. On the one hand, the explanation of the electron's spin, the explanation of the Zeeman effect, and the stunning and successful prediction of positrons indicate that the Dirac equation has put us on the right track. On the other hand, we have become convinced that a complete theory of electrons and positrons must take into account the possibility of electron-positron pairs being spontaneously produced. This pair production involves the electric field, and so a complete theory must include a quantum theory of electromagnetic processes as well.

The Dirac equation by itself doesn't give a coherent description of these processes for a general situation, although it works in a simple case like the electron-barrier situation. The problem is that quantum mechanics was really designed to deal with one particle at a time, so a process like pair production doesn't fit into the framework. This is bad news, because as we already saw, pair production is going on all the time. We need a new framework. The new framework, which harmonizes all three of our themes—quantum mechanics, special relativity, and field theory—is known as *relativistic quantum field theory* and is the basis for our modern theory of all particles and their interactions—the Standard Model. Except for gravity, of course.

You Can Always Get There from Here

It was 1942, and Dick Feynman was worried. He was A.B.D. at Princeton—All But Dissertation—and his advisor, John Archibald Wheeler, had disappeared to the University of Chicago where he worked for the mysterious Metallurgical Laboratory, which hadn't hired a single metallurgist. In fact, Wheeler was working with Enrico Fermi to produce a nuclear fission reactor. Feynman himself was already a part of the Manhattan Project, and had been working on a method to separate fissionable uranium from the useless, common variety. Feynman's fiancée had tuberculosis, which at the time was still incurable and usually fatal. And Feynman was worried about quantum mechanics.

When Feynman didn't understand something completely, he would tackle it from a completely different angle. For his Ph.D. thesis, he had decided to look for a new approach to quantum mechanics. He began by throwing out the Schrödinger equation, the wave function, everything everyone knew about quantum mechanics. He decided instead to think about the particles. He began with the two-slit experiment. It is as if the electron goes through both slits, he thought. Add the contributions for each path and square the result to get the probability that the electron will hit the screen at that point. What if we open a new slit? Then the electron "goes through" all three. If we add a second baffle with slits of its own, we need to add up all possible combinations of how the electron can go (only a few of which are shown):

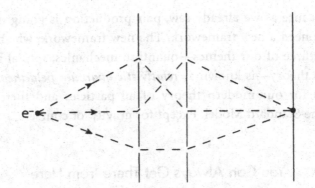

As we open up more slits in each baffle, there are more possible paths the electron can take, all of which we need to include in the sum. So, if we now remove the baffles, we should do an infinite sum over *all possible paths* the electron can take to get to that point on the screen. Of course, this includes paths that make a stop in Aruba, or on the moon!

Feynman learned of a suggestion Dirac had made for the contribution to the quantum field from a short segment of path. Amazingly, when Feynman combined this idea with his own sum-over-all-paths approach, he got the exact same answer as quantum mechanics, in every case. Feynman had succeeded in reformulating quantum mechanics in an entirely new framework, one that would prove crucial for relativistic quantum field theory. In 1942, he finished his Ph.D. thesis, in which he introduced this point of view, and then left for Los Alamos to work on the atomic bomb project. All of the best minds in physics were occupied with war work for the next few years and could only return to doing fundamental physics after the war ended. So it was not until 1946 that Feynman took up the idea from his thesis to see if he could make it work for a relativistic theory of electrons, namely, the Dirac equation.

In order to find the probability for some process to take place, Feynman knew he had to add up all the ways that it *could* take place. To keep track of all the possibilities, he started drawing little pictures for each possibility. The most basic action was for an electron to go from one place to another. Feynman represented this with a straight line between the two points, labeled by the symbol for an electron, e– (left in the figure below). As we know, the electron doesn't necessarily take this straight-line path. We need to take into account the contributions from all possible paths connecting the two points. The straight line is just a shorthand way of writing the sum of the contributions from all the paths. The only other basic possibility was for the electron to emit or absorb a photon, labeled by γ in the diagram:

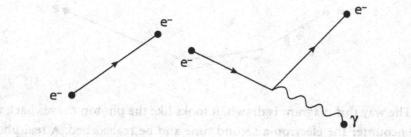

The squiggly line represents the photon. The electron line bends because the electron recoils when it emits the photon, like a fifth-grader throwing a medicine ball while standing on a slippery surface. In this way, Feynman could visualize the force between two electrons as due to an exchange of a photon:

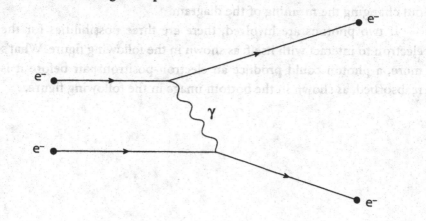

The first (emitting) electron recoils, like the child throwing the medicine ball. The second (absorbing) electron gets the energy and momentum of the photon like a second child who catches the ball. The result is that the two electrons swerve away from each other. So, the picture of electrons exchanging photons explains the fact that the electrons repel each other.

Feynman realized that in going from one point to another, an electron could interact with *itself* by way of the photon exchange. For instance, it could emit and then reabsorb a photon.

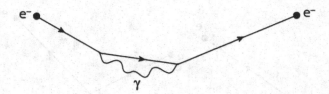

The way this diagram is drawn, it looks like the photon curves back to encounter the electron a second time and be reabsorbed. A real photon, of course, always moves in a straight line. But the photon in the diagram is a virtual photon, and the squiggly line doesn't represent its actual path. The virtual photon travels on all paths, just as the electron does. The only important points in the diagram are the interaction points. The lines can be drawn straight or curved as convenient without changing the meaning of the diagram.

If two photons are involved, there are three possibilities for the electron to interact with itself, as shown in the following figure. What's more, a photon could produce an electron-positron pair before it is reabsorbed, as shown in the bottom image in the following figure.

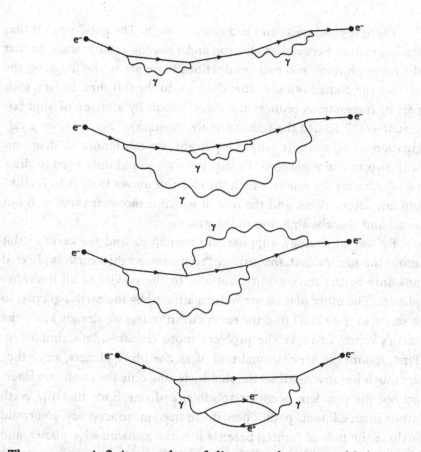

There are an infinite number of diagrams that we could draw, with three, four, or more photons, each of which can cause pair production…and we have to add them all up, just to find out how an electron gets from point A to point B. It seems we have an impossible task.

There is good news and bad news, though. The good news is that the interaction between an electron and a photon is fairly weak, so that the more photons and pair productions there are in the diagram, the smaller the contribution of that diagram to the full sum. In fact, each pair of interactions reduces the contribution by a factor of approximately 1/137 (called the fine-structure constant). Therefore, for a calculation to be accurate within 1 percent, we can ignore all diagrams with two or more photons. In that case, we would only need to draw two diagrams: the one in which the electron moves from A to B without any interactions, and the one in which it moves from A to B but emits and absorbs a photon in the process.

By way of analogy, suppose you wanted to find the earth's orbit about the sun. At first, the problem seems intractable: earth is affected not only by the sun's gravity, but also by the gravity of all the other planets. The other planets are in turn affected by the earth's gravity, so it seems as if we can't find the earth's orbit unless we already know the earth's orbit. To make the problem more tractable, let's simplify it. First, ignore the gravitational pull of all the other planets, since they are much less massive than the sun. Including only the earth-sun interaction, the problem becomes solvable, and one finds that the earth orbits in an elliptical path. Then, if you need more accuracy, you could include the pull of Jupiter, because it is the most massive planet, and you would get a slightly different orbit. For an even more accurate answer, you need to take into account the pull of the other planets, the pull of earth back on Jupiter, and so forth. Physicists call this a *perturbation expansion*—start with the largest influences, then add, one by one, the perturbing effects—in this case, the gravitational pull of the other planets. There are an infinite number of perturbations that need to be added, but fortunately they get smaller and smaller. This is what gives us hope for the theory of electrons and photons we are building. Because the fine structure constant is so small, we can start with the simplest diagrams and calculate an approximate answer. Then, if more accuracy is needed, we can add diagrams with more photons and electron-positron pairs until we reach the accuracy we need.

Now the bad news. When you actually calculate how much each diagram contributes, adding up all possible paths for the electrons and photons, you find that the diagrams with additional interactions in them, instead of giving a smaller contribution as expected, give an infinite contribution. This was not a surprise to Feynman; other physicists using other approaches had already discovered that you get infinite results for many of these calculations. In some cases, they had learned to circumvent the infinite results by doing the calculation in such a way that the infinite part of the answer was never needed. An example is the Dirac "sea" of negative-energy electrons. If you try to calculate the *total* energy, the answer is infinite. But if you only care about *changes* in the energy, for instance when an electron is ejected from the sea and a hole is formed, then the problem of the total energy can be ignored.

Feynman, however, had an advantage: Using his diagrams, he could classify all of the infinities that came up. It turns out that there are really only three fundamental infinities; all the other diagrams that give infinite answers are really manifestations of the three fundamental infinities. The three quantities that are infinite, according to the calculations, and in contradiction with actual experience, are the electron's mass, the photon's mass, and the electron's charge.

Before we investigate these infinities and learn how to eliminate them, let's take a closer look at what the theory is telling us about how electrons and photons behave. Do the electrons *really* take every path, including the one with a rest stop on the moon, to get to their destination? Suppose I release an electron and detect it three seconds later a short distance away. Wouldn't the electron need to travel faster than the speed of light to get to the moon and back in that time?

First of all, if we calculate the probability that we will actually detect our electron on the moon during the three-second interval, we get zero—absolutely no chance of detecting it there. Well, you say, that settles it; it can't ever be there, and so we can exclude all those faster-than-light trips from the calculation. But that won't work. If we exclude those paths, then everything goes haywire, and we get the wrong answer for the electron's actual trip. We need to add up *all* paths, not just the ones that seem reasonable, in order to get the right answer. The electrons engaged in bizarre behavior, such as traveling faster than light, are called *virtual* electrons, to distinguish them from electrons on physically reasonable paths, which we will refer to as *real* electrons.

Virtual particles may seem too weird to be relevant to the real world, but they are actually indispensable. For instance, take the simple problem of finding the force between two stationary electrons. We know they each feel a force from the other electron, because like charges repel. According to our theory, this force is caused by the exchange of photons. But how can a stationary electron emit a photon? If the electron recoils, it is no longer stationary. If there is no recoil, the laws of conservation of momentum and energy imply that the photon carries no momentum and no energy—a dud photon that can't affect anything.

In fact, the photons that are exchanged in this situation are all virtual photons. They live on borrowed momentum, borrowed energy, and borrowed time. They don't even travel at the speed of light! If you put a detector in the space between the electrons, you will never detect a photon. Because you can never detect them, they aren't considered real. If you like, you can consider them to be purely a calculational tool, artifacts of the way we calculate, without any actual existence. But they're such useful artifacts that physicists often talk about them as if they had an existence of their own.

QED

We now have a complete theory of electrons, positrons, and their interactions (except for the infinities mentioned earlier!). Physicists call this theory quantum electrodynamics, or QED for short—it is a quantum theory of dynamic (interacting) electrons. QED is our first example of a full-blown relativistic quantum field theory. Richard Feynman summarized the rules in his wonderful little book *QED: The Strange Theory of Light and Matter*. There are just three possible basic actions in the theory (and a picture corresponding to each action):

Action	Picture
An electron goes from place to place	
A photon goes from place to place	
An electron emits or absorbs a photon	

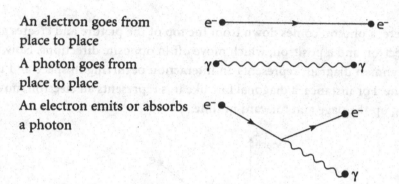

Together with these three basic actions, there is a rule for calculating: Using any combination of the three basic actions, draw all possible diagrams representing the different ways a process can happen. Calculate the value associated with each diagram. Add up the values for all the diagrams and square the result to get the probability that the given process will actually happen. Of course, the hard work comes in calculating the numerical value assigned to each diagram—you have to somehow do the sum over all possible electron and photon paths. Physics Ph.D. students spend years learning the mathematical techniques for computing these answers.

What about pair production? It might seem like we left that out when we listed the three basic actions of QED. Recall that a photon with energy $E = 2mc^2$ can produce two particles (an electron and a positron) with rest energy mc^2. In Feynman diagrams, pair production looks like this:

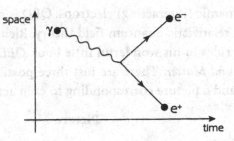

Here, a photon comes down from the top of the picture and creates an electron and a positron, which move off in opposite directions. Now, a Feynman diagram represents an interaction occurring in space and in time. For instance, a diagonal line like this represents an electron moving up the page and forward in time:

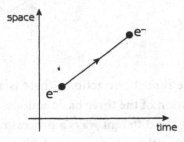

The vertical direction in the diagram represents space and the horizontal direction represents time. QED, of course, is a *relativistic* quantum field theory. Special relativity guarantees there will be an intimate connection between space and time in the theory. Because of this spacetime symmetry, we are allowed to take any Feynman diagram and rotate it by 90°. A 90° rotation exchanges the vertical and horizontal directions in the diagram; therefore, it interchanges the roles of space and time.

Take the basic interaction diagram (top, in the following figure), and rotate it 90°. In the rotated diagram (middle), we see a photon moving up the page on the left, and two electrons on the right—but one electron is moving backward in time. As we already know, an electron moving backward in time is equivalent to a positron moving forward in time. Make this switch, and we have exactly the diagram for pair production (bottom, in the figure): a photon enters and splits into an electron-positron pair. Because of the way space and time are interwoven in special relativity, we don't need to include this as a separate "basic action" in the rules for QED. Pair production is *automatically* included in the theory because of the interaction rule (the third "Action" in the earlier list) and the spacetime symmetry of special relativity.

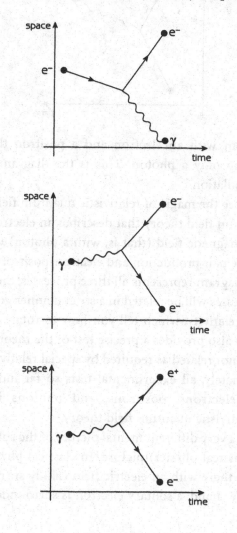

Electron-positron annihilation is automatically included, too. Just rotate the interaction diagram 90° the other way, and, after exchanging another backward-in-time electron for a positron, we get this process:

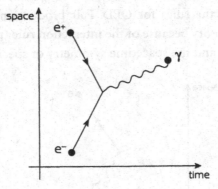

Here, we start out with an electron and a positron that meet and annihilate, leaving only a photon. This is the diagram for particle-antiparticle annihilation.

We begin to see the magic of relativistic quantum field theory. Any relativistic quantum field theory that describes an electron interacting with the electromagnetic field (that is, with a photon) will automatically also describe pair production and electron-positron annihilation. Since the same diagram represents all three processes, the probabilities of the three processes will be related in a strict manner governed by the rules of special relativity, which tell you how to rotate the diagrams. This relationship also provides a precise test of the theory: If the three probabilities are not related as required by special relativity, the theory is wrong. Fortunately, all experimental tests so far indicate that the interaction of electrons, positrons, and photons is beautifully described by relativistic quantum field theory.

QED creates a very different mental picture of the subatomic world than the old (classical physics) picture. In classical physics, a solitary electron just sits there with an electric field calmly surrounding it. In QED, on the other hand, a solitary electron is surrounded by a swarm of activity—virtual photons shoot out and back in; virtual electron-positron pairs are created and evaporate instantaneously. The "empty" space around the electron is abuzz with virtual particles. Let's think about the consequences of this model. When a virtual electron-positron pair is created near the (real) electron, the virtual positron will be attracted toward the real electron, while the virtual electron is repelled. There should be a resulting separation of charge: The real electron should be surrounded by a swarm of virtual positive charge. This charge shields the original negative charge, so that from far away, the electron should look like it has less electric charge than it actually has. Well, we can calculate by how much its true charge is reduced, using the rules of QED, and we find the answer is infinite. This is precisely one of the three fundamental infinities of QED that were mentioned earlier. In order to deal with the infinite amount of shielding charge, we have to assume that the bare (unshielded) electron has *infinite* negative charge, so that we can add an infinite amount of positive shielding charge and still end up with the small, finite, negative charge that we know the electron actually has. This technique is called *renormalization.* You redefine the electron's charge in order to make the result of the calculation agree with the charge you actually measure.

Renormalization amounts to subtracting infinity from infinity and getting a finite number. According to the mathematicians, "infinity minus infinity" is meaningless. Physicists are not as picky about such niceties as are mathematicians. Still, they too felt that the situation was unsatisfactory—an indication that there was something wrong with the theory. But, being physicists and not mathematicians, they went on doing it as long as it worked, and ignored the contemptuous glares of their mathematical colleagues. It has continued to work, though, so successfully that now renormalizability is considered the hallmark of a viable relativistic quantum field theory. The other infinities in QED, the electron mass and the photon mass, can be dealt with in the same way. Once this is done with the fundamental infinities, all the Feynman diagrams give finite results—not only finite, but astoundingly accurate.

To test the accuracy of the renormalization procedure, imagine trying to penetrate the shielding cloud of virtual particles. By firing higher and higher energy particles at the electron, it should be possible to penetrate the cloud of positive charge, getting closer to the bare, unshielded, charge of the electron. So we expect that, as we increase the energy of the probe particle, the apparent electric charge of the electron will increase. This is exactly what happens in accelerator experiments! At the extremely high energies needed to produce Z^0 particles (about which we will learn a great deal in Chapter 9), the effective electric charge is 3.5 percent higher than it is in low-energy experiments, in good agreement with the theoretical prediction. (Actually, only half of this change is due to virtual electron-positron pairs. The other half is due to virtual particle-antiparticle pairs of other particle types.) This gives us confidence in the renormalization procedure in spite of its questionable mathematical status.

Some physicists, notably Dirac himself, were never happy with the renormalization procedure, and considered it a way of sweeping the problems under the rug. Steven Weinberg, one of the founders of the Standard Model, feels differently. He points out that the electron's charge would be shielded by the cloud of virtual particles even if there were no infinities in the calculations, and so renormalization is something we would have to do anyway. It is possible, though, that the theory itself is giving us hints about where it breaks down. The infinities come from the virtual particles with very high energy. Maybe QED is simply incorrect at very high energies. Is there a theory that "looks like" QED at low energies, but at high energies is different enough that no infinities arise? In the past 20 years, some physicists have come to suspect that this is possible, and even have an idea what the underlying theory might look like. We'll take a look at those ideas later on in the book.

Chapter 6

Feynman's Particles, Schwinger's Fields

Scientific truth should be presented in different forms, and
should be regarded as equally scientific, whether it appears in
the robust form and the vivid colouring of a physical illustration, or in the tenuity and paleness of a symbolic expression.
—James Clerk Maxwell,
address to the British Association, 15 September 1870

Quantum electrodynamics (QED) is our first example of a complete relativistic quantum field theory. It only describes electrons,
positrons, and photons, leaving out most of what makes up normal
matter—protons and neutrons. Nonetheless, QED has the characteristics that all relativistic quantum field theories share: particle-antiparticle
symmetry, forces carried by intermediate particles, Feynman diagrams,
sum-over-paths, renormalization, shielding of charges, and the perturbation expansion.

As the name implies, relativistic quantum field theories are born of
quantum mechanics, and from it, they inherit both wave and particle
aspects. The previous chapter described QED in terms of particles:
electrons, positrons, and photons, and interactions that are possible for
those particles. What does QED look like when we describe it in terms
of quantum fields?

The Best of All Possible Worlds

There are two ways of doing classical (that is, prequantum) physics. One possibility is to specify the state of the universe (or some part of it) *now* and give the rules for how to get from now to *now-plus-a-little-bit*. That is, we say where each particle is, where it's going, and the forces that influence its motion. Let's call this the local approach to physics, because what happens to a particle next depends only on influences (other particles, fields) in the particle's immediate neighborhood. What we're doing is taking a snapshot of the universe at a particular time, and using that snapshot, plus the laws of physics, to predict what will be happening at a slightly later time.

The other possibility is to look at where the particle (or particles) starts, where it ends up, and say that somehow what it does in between is optimal; it follows the best path, in some sense. Let's call this the global approach.

For instance, the light scattered off a fish in an aquarium bends (refracts) when it leaves the tank and enters the air. In the local point of view, a ray of light that leaves the fish simply goes straight ahead as along as it is still in water, according to Maxwell's equations for light. Then, when the water meets the air, another application of Maxwell's equations tells us that the ray takes a turn, which depends on the ratio of the speed of light in water versus in air. (This is known as *Snell's law*.) Finally, the ray again travels in a straight line until it enters your eye.

In the global approach, we look at the entire path from fish to eye, and ask: What is the optimal path? The answer, as discovered by Pierre de Fermat in 1661, is surprisingly simple: The path actually taken by the light ray is the one that takes the least time to get to the eye.

To understand this, think about a lifeguard at the seashore who wants to save a floundering swimmer. He knows he can run faster than he can swim, so the straight-line path is not the fastest, as he would spend too much time in the water.

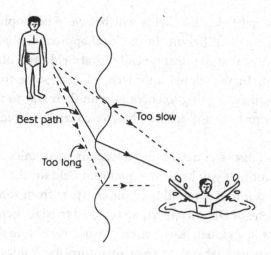

Instead, he should run along the beach toward the water, then turn and start swimming. The path of least time is given by the same law as in the case of the light ray in the aquarium.

There is a variation of this least time principle that works for particles: It's called the *least action principle*. According to this principle, we calculate the difference between the kinetic energy and the potential energy (this difference is called the *Lagrangian*) for each point along the path. Then add the Lagrangian values for the entire path. This sum gives the *action* for the path. The least action principle declares that the path actually taken by the particle is the path with the smallest action. In the global approach we look at the whole path from beginning to end, not just what's happening at one instant. Then, from among all possible paths, we choose the optimal path.

It turns out that the local and global approaches are mathematically identical. Think of the Lagrangian as something that summarizes all the equations of the theory. The least action principle tells how to extract those equations from the Lagrangian. By a careful choice of the Lagrangian, we end up with exactly the same equations as in the local approach. The two approaches, therefore, give the same predictions about how particles and fields will behave. Philosophically, however, they seem totally different. In the local approach, objects only "know about" things that are nearby. Only local fields and objects can influence them. In the global approach, objects seem to "know about" everything they're going to encounter on their way to the finish line— they somehow see the whole obstacle course and pick the best path through it.

Schrödinger's equation for quantum mechanics was in the best local tradition. If you know the quantum field for the electron at some time, you can find it for all later times. Apart from some cryptic comments in one of Dirac's papers, no one had tried to formulate quantum mechanics in a global, least-action perspective before Feynman's Ph.D. thesis. Feynman set out to treat quantum mechanics using the least action principle, and this led him to the radically different view of particles in his sum-over-paths approach. We can already see a connection: In the classical physics case, the particle "looks ahead" and chooses the path with the smallest action; for Feynman, a quantum particle looks ahead on all possible paths, and decides its probability of being in a certain place based on the result of the sum over all paths.

Imagine the Lagrangian forming a trough through which the particle moves:

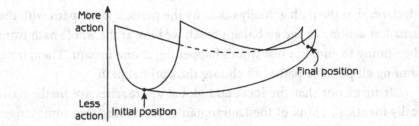

In classical physics, the particle stays in the very bottom of the trough, in order to end up with the smallest possible action. In quantum mechanics (and also in quantum field theory), the particle follows all paths, including ones that go high up on the sides of the trough and zigzag all over the place. Now, a particle's quantum field can be positive or negative (to be precise, it is a complex number, but that doesn't matter as far as we are concerned). As the particle moves along a path, the quantum field oscillates from positive to negative. This is called the phase of the quantum field: whether and by how much it is positive or negative. According to Feynman, the oscillation of the phase is controlled by the Lagrangian, the same quantity that appears in the least action principle. The steeper the slope of the trough, the faster the phase of the particle oscillates. As a result, very small differences in paths on the walls of the trough cause large differences in phase. Then, when we add up all the paths, those on the walls tend to cancel out, because for every path with a positive phase there is a nearby path with a negative phase. In contrast, in the bottom of the trough, phases don't change very fast, so all nearby paths have positive phase, or else they all have negative phase. Instead of canceling out, they reinforce each other. It is rather like gossip: the more directly it is transmitted to you, the less likely it is to vary from the original source. Gossip that takes a round-about path through 15 or 16 people is unlikely to be reliable, and in fact, may contradict what you're hearing from other sources. In that case, you get a cancellation: Since you are hearing contradictory stories, you don't believe either version.

Now we can ask again a question we have only touched on so far: If I'm made of particles, and my particles behave in such bizarre ways, traveling on every possible path, winking in and out of existence, then why is it that I don't also behave that way? Why can't I stop at the bank and the school and the post office simultaneously on my way to work? The answer lies in the fact that the larger an object is, the faster its phase oscillates. My body is made of something like 10^{28} protons, neutrons, and electrons, so my phase oscillates faster, by a factor of about 10^{28}, than the phase of a single proton. As a result, all paths but the least action path experience cancellations from nearby paths. To say it another way, the trough for me, or for a rock, or a chair, is so narrow and deep that I have to stay at the bottom of it.

From Feynman's approach, then, we can see how the world can look quantum mechanical when we look at small particles, but still look classical for everyday objects. According to Feynman, we don't live in "the best of all possible worlds"—we actually live in *all* possible worlds. But most of these worlds cancel each other out, leaving, for large objects like ourselves, only the "best" world, the one with the smallest action.

Feynman's diagrammatic approach was tremendously efficient for doing calculations. Feynman, however, was not the first person to calculate an experimentally testable number using relativistic quantum field theory. That honor went to another brilliant young theorist, Julian Schwinger.

"Schwinger was to physics what Mozart was to music," according to one of his colleagues.[1] He started college at the age of 16 and wrote his Ph.D. thesis before he got his bachelor's degree. In personality, he was a complete contrast to Feynman. Feynman was a casual dresser, a practical joker, and outspoken. Schwinger was shy and introverted. He always worked in his jacket and tie, never loosening the tie in the hottest weather. He would show up at the physics department sometime in the afternoon, have a breakfast of "steak, French fries and chocolate ice cream," according to a friend, and start work at 7 or 8 P.M.

"He would be leaving for home in the morning when other members of the department were arriving."[2] Where Feynman thought in pictures, Schwinger thought in equations. Rather than developing radically new ways to think about a problem, Schwinger was a genius at taking a known formulation and developing it to its fullest extent. His writing style was dense; to some, impenetrable. "Other people publish to show you how to do it, but Julian Schwinger publishes to show you that only he can do it," said one critic.[3] In fact, many later developments of quantum field theory by other people were discovered buried in Schwinger's earlier published papers. No one at the time had understood what he was doing well enough to make use of it.

Schwinger's approach to QED was from the field theory point of view. In Chapter 1 we learned that a classical field is simply an arrow at each point of space. A quantum field, according to Schwinger, can be pictured as a quantum harmonic oscillator at each point in space. We know what a quantum harmonic oscillator looks like: a bowl with equally spaced energy levels. When there are no particles anywhere (a state called the *vacuum*), all of the oscillators are in their lowest energy level. A single particle at one point in space is represented by moving the oscillator at that point up one notch to the next higher energy level:

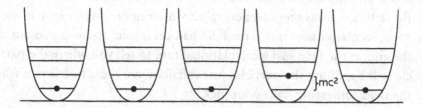

The difference in energy is exactly mc^2, the rest energy of a particle (or quantum) of the field. Bumping the oscillator to the next higher energy level adds an equal amount of energy; another mc^2. Therefore, this level represents two particles at that point in space. (For fermions—like the electron, for example—the Pauli exclusion principle prohibits two particles in one spot. For bosons, however, there is no difficulty in having any number of them in one place.) To represent a moving particle, imagine the energy leaping from one oscillator to the next. One oscillator drops down in energy just as a neighboring oscillator jumps up in energy.

In 1949, Freeman Dyson, who understood both Schwinger's field theory point of view and Feynman's sum-over-paths approach, proved that the two approaches were actually identical. It was the same mathematics, the same theory, just two different interpretations of the mathematics. For physicists, this meant they could use the Feynman diagrams that made calculations so convenient, and fall back on Schwinger's rigorous approach when there were ambiguities, or when the powerful techniques of field theory were called for. For philosophers, it reinforced the answer that quantum mechanics gave to the old question: Wave or particle? The answer: Both, and neither. You could think of the electron or the photon as a particle, but only if you were willing to let particles behave in the bizarre way described by Feynman: going all ways at once, spontaneously popping into existence and disappearing again, interfering with each other and canceling out. You could also think of it as a field, or wave, but you had to remember that the detector always registers one electron, or none—never half an electron, no matter how much the field had been split up or spread out. In the end, is the field just a calculational tool to tell you where the particle will be, or are the particles just calculational tools to tell you what the field values are? Take your pick.

The Evidence

QED survives, nearly unchanged, as a subset of the Standard Model. Other parts of the Standard Model are harder to test, but the QED part has been subjected to many tests and has never yet failed. The first experimentally verified result of the theory, published by Schwinger in 1948, has been called the most accurate prediction of any scientific theory, namely the magnetic properties of the electron. An electron, we know, has a spin like a top. In empty space, a top would always point in the same direction, but here on earth, a top turns or precesses as it spins. An electron, we know, behaves like a tiny magnet. Now, a magnet, when placed in an external magnetic field, precesses like a top.

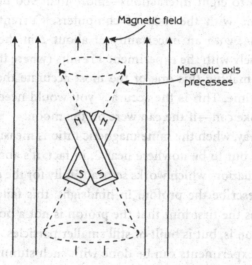

Magnetic field

Magnetic axis precesses

If the electron were a tiny spinning ball of charge, Maxwell's equations would let us find the effective magnetic strength, known as the *magnetic moment*, from the spin and the charge of the electron. But we already know that we can't think of the electron that way—the spinning ball picture of the electron creates too many problems. Fortunately, the Dirac equation comes to our rescue. The Dirac equation predicts that an electron will precess exactly twice as fast as the spinning ball picture predicts. The fact that the Dirac equation got the factor of 2 correct was a tremendous success, and helped enthrone that equation as the correct description of the electron. By 1947, however, "the sacred Dirac theory was breaking down all over the place," as

Schwinger put it.[4] New, more precise experiments revealed that instead of a factor of 2, the ratio was more like 2.002. This small (only one-tenth of one percent) but significant difference was the main impetus behind Schwinger's and Feynman's development of QED. Schwinger won the race. The cloud of virtual particles around the electron, like a cloud of gnats around a hiker, altered the motion of the electron by just the right amount. Today, more precise calculations include all Feynman diagrams up to eight interactions—more than 900 diagrams in all—and are done with the help of computers. Currently, the value is 2.0023193044, with an uncertainty of about 2 in the last digit. This agrees precisely with the experimental results (where the uncertainty is much less), making this one of the most accurate theoretical predictions of all time. This is the accuracy you would need to shoot a gun and hit a Coke can—if the can were on the moon!

By the way, when the same magnetic ratio is measured for the proton, it turns out to be nowhere near 2; in fact, it's about 5.58. Clearly, the Dirac equation, which works so beautifully for the electron, is inadequate to describe the proton. In hindsight, this failure of the Dirac equation was the first hint that the proton is not a point-like particle, as the electron is, but is built of still smaller particles.

Not all experiments can be done with such stunning one-part-in-a-billion precision. Many other aspects of QED have been tested experimentally, however. The increase of mass of particles in accelerators (in accordance with special relativity); the probability a photon will produce an electron-positron pair; the probability a photon will scatter in any given direction off an electron; the detailed explanation of the spectra of atoms; all can be calculated in QED and all have been confirmed (with appropriate allowances for phenomena involving particles other than electrons, positrons, and photons). What's more, QED has continued to be accurate at smaller and smaller distances. As physicist Robert Serber has pointed out, "Quantum mechanics was created to deal with the atom, which is a scale of 10^{-8} centimeters," or one hundred millionth of a centimeter. Then, it was applied to the nucleus of the atom, a hundred thousand times smaller, "and quantum mechanics still worked.... Then after the war, they began building the big machines, the particle accelerators, and they got down to 10^{-14}, 10^{-15}, 10^{-16} centimeters, and it still worked. That's an *amazing* extrapolation—a factor of a hundred million!"[5] Not bad for a theory that was thought throughout most of the 1930s and 1940s to be so "crazy" and "ugly" (in Victor Weisskopf's words)[6] that it wasn't worth working on.

Can we really believe a theory that is this crazy? After all, my body is made of protons, neutrons, and electrons. Am I to believe that every time I walk from the couch to the refrigerator, my electrons make virtual trips to Hawaii, the Bahamas, and Mars? As Richard Feynman put it:

> It is not a question of whether a theory is philosophically delightful, or easy to understand, or perfectly reasonable from the point of view of common sense. The theory of quantum electrodynamics describes nature as absurd from the point of view of commonsense. And it agrees fully with experiment. So I hope you can accept nature as she is—absurd.[7]

The Great Synthesis

The full theory of QED finally came together in 1948 at an amazing conference of the top theoretical physicists that met at Pocono Manor Inn in Pennsylvania. Many of the participants had been doing applied work during the war years, such as the atomic bomb project, and they were eager to return to real physics, fundamental physics. Schwinger gave a five-hour talk in which he showed how all of the infinities that kept cropping up in calculations could be removed. His methods, convoluted but rigorous and precise, left heads spinning. Then, Feynman got his chance. He introduced his diagrams along with other unconventional techniques, he made arbitrary assumptions, he invented rules out of thin air when he got into difficulty, and waved away the audience's objections. He got the same results as Schwinger, but he convinced no one that his grab bag of techniques would work. After the conference, Robert Oppenheimer received a letter from a Japanese physicist, Sin-itiro Tomonaga, who had independently come to the same conclusion as Schwinger and Feynman. All the infinities could be avoided; QED was a consistent and usable theory. Feynman, Schwinger, and Tomonaga received the Nobel Prize in physics in 1965.

QED describes only the interactions of electrons, positrons, and photons. It has nothing to say about nuclear forces, or about the multitude of new particles that started showing up in experiments. For the first time, though, physicists had a theory that brought together quantum mechanics and special relativity. QED was the first complete relativistic quantum field theory. Other theories would follow as physicists tried to understand the nuclear forces and the new particles, but again and again relativistic quantum field theory provided the theoretical structure underpinning the solution.

In nineteenth-century physics, the universe contained two things: particles and fields. Particles were tiny and hard, like small billiard balls. Fields were elastic and spread throughout space. Particles produced fields according to their electric charge and their motion, and particles responded to the fields of the particles around them. Relativistic quantum field theory completely eliminates the distinction between particles and fields. Matter (the electrons and positrons) and forces (the photons) are both described in the same way—by quantum fields. Quantum fields combine particle nature and field nature in a single entity. The quantum field spreads out through space, just like a classical field, but it is *quantized*: When you try to measure the field, you always find a whole particle, or two particles, or none. You never detect half an electron or a fractional photon. If fields were peanut butter, classical fields would be smooth and quantum fields would be chunky.

This dual nature of quantum fields is reflected in the two descriptions physicists employ when working with them. Feynman diagrams picture the field as a dense mesh of tiny interacting particles. Schwinger instead pictured the field as a continuous collection of harmonic oscillators, spread throughout the universe. Both pictures are useful in their own way. Schwinger's field picture makes the symmetries of the theory apparent and emphasizes the continuous nature of the quantum field. Feynman diagrams are easy to visualize, they emphasize the particle aspect of the theory, and they greatly simplify the process of calculation. We will need to call both pictures into play in the coming chapters.

Special relativity is built into the structure of relativistic quantum field theory in a fundamental way. We are thus guaranteed that (real, rather than virtual) massless particles will always travel at the speed of light, and that particles with mass will never exceed it. The twists and flips we can do with Feynman diagrams, the fact that an electron moving backward in time is identical to a positron, and the ability to create matter and antimatter from pure (photon) energy, these are all consequences of the intimate connection with special relativity.

Relativistic quantum field theory harmonizes the formerly discordant strains of fields, relativity, and quantum mechanics. In achieving that harmony, physicists had to throw out much of what they thought they knew. Physicists began, back in the days of Galileo, by describing the behavior of everyday objects, falling rocks and rolling spheres. As they moved farther from everyday experience into the world of atoms and molecules, electrons, protons, and neutrons, they carried the tools they had developed: the concept of a particle as an ideal mathematical point, and the concept of a wave in an ideal continuous medium. In hindsight, it is perhaps not so surprising that these concepts, abstracted from experience with large objects, should fail when confronted with the world of the very small. The new concept, the quantum field, was an attempt to describe a phenomenon so far beyond our experience that analogies to everyday objects simply didn't work anymore. The microworld was a much stranger place than anyone could have imagined. Even physicists trained in the mathematical techniques of field theory need help to develop physical intuition about this new world. Here the two sets of mental images, field and particle, are of immeasurable assistance.

Relativistic quantum field theory is a milestone in intellectual history: a theory that goes beyond what can be represented in simple pictures and analogies, where the mathematics must be the guide and the mental images of wave and particle are of secondary importance. The images can help shape a physicist's intuition, but the mathematics is the ultimate arbiter. Relativistic quantum field theory finally answered the long-standing question of the nature of light: Is it particle or wave? The answer: Light was something completely new—a quantum field, neither particle nor wave, but with aspects of both. But the revolution in thought goes deeper than that. Matter, too, can be described by a quantum field. Electrons are just as ineffable and evanescent as photons. If electrons, then why not protons and neutrons, too? Could all matter be described by quantum fields? Was the world of solid objects going to dissolve into an amorphous sea of probabilities and virtual particles? The answers would only come with the development of the Standard Model.

Chapter 7

Welcome to the Subatomic Zoo

The changing of bodies into light, and light into bodies, is
very conformable to the course of Nature, which seems
delighted with such transmutations.
—Isaac Newton, *Opticks*

How do you find out what something is made out of if you aren't
allowed (or able) to open it and look inside? Imagine you are given
what looks like a Nerf ball. Your mission (should you choose to accept it)
is to find out what's inside, without squeezing it or cutting it open. The
only things you are allowed to do are weigh it and shoot BBs at it. When
you weigh it, you find it weighs too much to be Nerf foam all the way
through—it weighs as much as a baseball. Next, you buy a BB gun and
start shooting at the ball. You find that most of your BBs get stuck inside,
and the ones that scatter off don't form any kind of informative pattern.
You go back to the store and buy a more powerful BB gun. Now you start
to see regular patterns in the BBs that fly off the ball.

It's time to stop and think. What are the possibilities? Your ball may
be an actual baseball, covered with a thin layer of Nerf foam just to
confuse you. What sort of pattern of BBs would you expect to see if
that were the case? At very low power, BBs would penetrate the Nerf
layer but bounce off the baseball's outer skin. At higher power, most of
the BBs would penetrate the baseball and get stuck there. At very high
power, the BBs would rip straight through, perhaps being deflected to
one side or another in the process. Another possibility is that, instead
of a baseball, there's a much denser object embedded in the Nerf, say, a
steel ball. In this case, we would expect that at low power the BBs get
stuck or slightly scattered. At high power, they would go right through

the Nerf and scatter off the hard center. In fact, by looking at the paths of the BBs that fly off and tracing them backward we can find out how large the central object is. There are still other possibilities we need to consider. What if there was more than one hard object embedded in the ball? Scattering would be more confusing. There would still be a low-power region where BBs mostly get stuck. At higher power, the BBs could bounce off one of the hard objects, or it could ricochet between several of them before exiting the Nerf and flying off into the distance. The patterns would be harder to identify. Still, we can imagine that with enough BBs and a little ingenuity, we could deduce the number and size of the hard objects.

This is, more or less, the situation that the elementary particle physicist faces. Back in 1909, Ernest Rutherford had pioneered this approach by firing alpha particles (which we now know to be helium nuclei—two protons and two neutrons) at a thin gold foil. The scattering pattern surprised him. From what he knew about the density of gold and the thickness of the foil, he had expected the alpha particles to pass through the foil with only a small amount of deflection. Instead, he found that some of the particles came right back toward the source, as if they had bounced off of a brick wall. "It was quite the most incredible event that has ever happened to me in my life," he said. "It was almost as incredible as if you fired a 15-inch shell at a piece of tissue paper and it came back and hit you."[1] In order to explain this unexpected result, Rutherford proposed that the mass of the gold atom wasn't evenly spread throughout the volume of the atom. Rather, the atom had a small, hard core that held 99.95% of the mass of the atom. This is like the second model in the Nerf example; the hard core is the nucleus of the atom, composed of protons and neutrons, and the remaining 0.05%, the Nerf foam, is the electron cloud.

This model is similar to the solar system, with the nucleus where the sun would be. The sun contains 99.9% of the mass of the solar system. Instead of the electrons being in orderly planet-like orbits, however, they must be spread out in orbitals as required by quantum mechanics (and by QED). It would be another 50 years before physicists would take the next step and probe inside the proton and neutron to discover their structure. As we will see, that structure is something like the third model in the Nerf example; three hard objects called quarks, and a cloud of "glue" that holds them together.

The process of firing particles at other particles and looking at what flies out has been compared to taking two Swiss watches and smashing them together, then trying to deduce how they work by looking at the gears and jewels that fly out. The actual situation is both better and worse than that. On the one hand, the proton structure is much simpler than a watch. It is built of just three quarks, plus the glue that binds them. On the other hand, we've never actually detected any of the "gears"—the quarks. Instead, when you smash two "Swiss watches" together—two protons, say—you find that two alarm clocks and a grandfather clock fly out. But let's go back and tell the tale from the beginning.

Our modern understanding of the nuclear forces grew via a complex process of development spanning about 40 years. In order to keep track of the main ideas through the twists and turns of that history, it may be useful to keep in mind the following outline. The history of the "strong nuclear force" can, in retrospect, be broken into three stages:

- Isospin Theory, 1930–1960
- The Eightfold Way, 1961–1973
- QCD, 1974–present

A New Force of Nature

It was only in 1662 that the ancient Greek atomic hypothesis was revived by Robert Boyle. Boyle noticed that a gas, such as air, could be compressed, which implied there was some empty space in between the parts making up the gas. In the early 1800s, Joseph-Louis Proust, John

Dalton, and other early chemists made use of the atomic hypothesis to explain why chemical reactions always occurred in fixed proportions. For instance, if you take one gram of hydrogen and 8 grams of oxygen, it will combine completely to make 9 grams of water. This can be explained by supposing that one atom of hydrogen combines with one atom of oxygen, and that each atom of oxygen is 8 times heavier than a hydrogen atom. This clearly is not the only solution: We now know that oxygen is 16 times heavier than hydrogen, and that two hydrogen atoms combine with each oxygen atom to make water (H_2O). It took much work for chemists to sort through the possibilities, but by 1869 Mendeleev had a reasonably good chart of atomic weights—the first version of what we now call the periodic table of the elements.

By 1935, physicists had an explanation for Mendeleev's periodic table. The "uncuttable" atoms had been cut and found to consist of protons, neutrons, and electrons. The electrons were bound to the nucleus by the electric force, operating according to the laws of quantum mechanics (later quantum electrodynamics, or QED). The only remaining question: What holds the nucleus together? The protons, positively charged particles, were crammed into a space $1/100,000^{th}$ the size of an atom. Positively charged particles repel each other, so the nucleus would fly apart if there was no other force holding it together. If only they could nail down the nature of the nuclear force, physicists thought, they would have a complete and fundamental understanding of matter.

"Science," said Robert Millikan in his Nobel acceptance speech in 1924, "walks forward on two feet, namely, theory and experiment. Sometimes it is one foot which is put forward first, sometimes the other, but continuous progress is only made by the use of both." In 1931, Dirac had stunned the physics community by predicting the existence of a new particle, the positron, which the experimenters then confirmed. Theorists came up with two more surprises during the 1930s.

The first surprise was triggered by the need to explain an experimental result that seemed to violate nearly every law of physics. The phenomenon is called *beta decay*; it is a form of radioactivity in which the nucleus emits an electron (then known as a beta ray) and changes to a different element. The simplest example is the decay of a free neutron (that is, one not bound in a nucleus). This free neutron gets converted to a proton, and an electron is emitted. Now, electric charge is conserved in this reaction. The neutron has 0 charge before the decay, and the proton and electron have +1 and −1 charge, so afterward the total charge is still 0. But everything else is screwy. For instance, spin. A neutron has spin 1/2, like the proton and the electron. But there is no way for initial spin to equal final spin—we can add two spin 1/2 particles to get spin 0 (by having one spin-up and one spin-down) or spin one or minus one (both spins in the same direction). But there is no combination that gives spin 1/2 for the total spin. So angular momentum apparently isn't conserved.

Energy apparently isn't conserved, either. A neutron is more massive than a proton and an electron combined. The extra mass should show up as kinetic energy (remember $E = mc^2$) of the proton and electron. Think of this extra energy as the gunpowder in a rifle cartridge. If two cartridges have the same amount of gunpowder, the two bullets will come out of the gun with the same speed, and therefore the same energy. Since the mass difference is always the same in any beta decay, the electron should always come out with the same energy. Instead, experiments found that the electron's energy was spread over a wide range. Initial energy, it seemed, did not always equal final energy.

Some physicists took this to mean that the sacred law of energy conservation had broken down. Wolfgang Pauli thought of a solution that seemed only slightly less radical—that a new particle, with spin 1/2 and no electric charge, was produced in the decay. How little confidence he had in this scheme is shown by the fact that he mentioned the idea in a letter to friends (he was explaining that he would not be at an upcoming physics conference because he wanted to enter a dance contest happening at the same time) and then let the matter lie. Several years later, in 1934, Enrico Fermi picked up the idea and published a paper showing how the new particle, dubbed the *neutrino* (or little neutral one), explained the range of energies seen in beta decay experiments. (For reasons we will learn later, physicists decided to re-name the new particle the *antineutrino*, and call its antiparticle the neutrino.) We will talk more about beta decay in Chapter 9.

The second theoretical surprise came from Hideki Yukawa in 1935. Yukawa had been struggling with the nature of the nuclear force holding the protons together. Several years before, Heisenberg had suggested that the nucleus was held together by an exchange of some particle among the protons and neutrons, rather like the electric force is carried by exchange of (virtual) photons. Yukawa realized, as Heisenberg had, that the particle being exchanged could not be an electron, for the same reason as in beta decay—the spins wouldn't add up. Finally, after hearing about Fermi's theory of beta decay and the invention of a new particle, Yukawa decided to assume the nuclear force was caused by the exchange of some particle, which came to be called the pion (pronounced "pie-on," as in the Greek letter pi), and then figure out just what the required properties of that particle would be.

In beta decay, a neutron turns into a proton, and an electron and an antineutrino are produced. Yukawa began with the same basic idea—namely, that a neutron could turn into a proton by emitting a negatively charged particle. The inverse reaction must also be possible: a proton turns into a neutron and emits a positively charged particle. That meant that the pion must come in two versions, a negative pion (called the pi-minus) and a positive pion (the pi-plus). Because the proton and neutron both have spin 1/2, the pion (in both versions) must have spin 0. The mass of this new particle can be estimated very simply. In Yukawa's theory, the nucleus is held together by virtual pions in the same way that electrons are bound to atoms by virtual photons. For this to work, the virtual pions must live for about 7×10^{-24} second—

the time it would take a particle traveling at the speed of light to cross the nucleus. Recall, though, that virtual particles live on borrowed time: the time they can survive is limited by the Heisenberg uncertainty principle, which states that the uncertainty in energy and time are related by the equation $\Delta E \, \Delta t \geq \hbar$. Because the pion has some mass m, it needs to borrow energy $E = mc^2$. Using the time 7×10^{-24} second for Δt in the uncertainty principle, we find that the pion mass should be about 200 times the electron mass. Yukawa rather hesitantly suggested that such a particle, with charge $+1$ or -1, no spin, and mass intermediate between an electron's mass and a proton's mass, should be sought in cosmic ray showers.

On a Mountaintop

As theorists struggled with the seemingly intractable problems of beta decay, the nuclear force, and the infinities in quantum field theory, experimenters were going up in balloons and climbing mountains. Their purpose was not recreational; they were trying to understand the nature of cosmic rays. By 1935, it was known that these mysterious rays came from the sky, consisted of charged particles, and were capable of penetrating several feet of lead. It was suspected (correctly) that, whatever the primary rays from space were, they interacted in the atmosphere to create many different types of particles.

Experiment again moved to the forefront of physics. Its ascendancy would last through the 1940s, 1950s, and 1960s, as a whole zoo of subatomic particles was discovered using a variety of increasingly sophisticated techniques. The bewildering array of exotic, short-lived particles would only find theoretical explanation in the 1970s with the development of the Standard Model.

It was Carl Anderson, the famed discoverer of the positron, who in 1937 won a tight race to claim discovery of a charged particle with mass, "larger than that of a free electron and much smaller than that of a proton," in fact, about 200 times the electron's mass. People assumed immediately that this was Yukawa's pion—but they were wrong. As the carriers of the nuclear force, pions ought to behave dramatically differently when traveling through matter than when traveling through empty space: They should be quickly trapped by a nucleus, which should then break apart in a shower of neutrons and protons. Two young Italians, working in a basement with scavenged supplies, decided to test if the new cosmic ray particles had the right properties. It was 1944, and Rome was occupied by the Germans. The city was falling apart around them. Still, Marcello Conversi and Oreste Piccioni managed to prove that these new particles lived more than 2 microseconds (2×10^{-6} seconds), much too long to be Yukawa's pions.

The new particles were found to have spin 1/2, and didn't seem to interact strongly with anything. Any particle that was produced in the upper atmosphere and that interacted strongly with the nucleus would have plenty of opportunity to do so on its trip down to sea level, as it would pass the nuclei of many air molecules on the way. It would be scattered many times and eventually captured by a nucleus. Therefore, it would have scant probability of ever reaching sea level. Because the new particles were detected in abundance at sea level after passing through miles of atmosphere, they couldn't have a strong nuclear interaction. After the war, Conversi and Piccioni teamed up with Ettore Pancini and conducted an experiment that settled the matter. They decided to look for the expected nuclear interactions by measuring how quickly the particles were absorbed in carbon rods. Instead of being rapidly absorbed by the carbon rods, the new particles decayed in the same way as in the atmosphere. This meant that their rate of capture by the carbon nuclei was a trillion times slower than that predicted for Yukawa's pions.

The new particle had the wrong spin, the wrong lifetime, and didn't interact strongly with the nucleus. It wasn't Yukawa's pion—that much was certain. It came to be called the muon (denoted by the Greek letter mu, μ), and physicists came to think of it as a sort of overweight older brother for the electron. Here was a coup for the experimenters and a curve ball for the theorists. No one knew why there should be another particle just like the electron, but heavier. The curmudgeonly I. I. Rabi groused, "Who ordered *that*?"

Through the end of the 1940s and into the 1950s, cloud chambers continued to be built at higher and higher altitudes to take advantage of the greater flux of cosmic rays. Particle physics became a sort of extreme adventure sport: long days in cramped quarters while the storms blew outside, skiing on the glaciers in good weather, and, not infrequently, death in the crevasses. Detection methods were improved. Geiger counters were added to the cloud chambers, so that the chamber would be triggered and a photograph taken only when a particle was detected by the Geiger counter. A new technique was developed using photographic emulsion, essentially a type of camera film with an extra-thick layer of chemicals that could be left on the mountains for months at a time, then developed and examined microscopically for particle tracks. With this method, a track as short as a millionth of a centimeter could be seen.

With the new techniques, experimenters kept coming up with new surprises. A particle was discovered with spin 0, and mass about 270 times the electron mass, which would occasionally be seen colliding with a nucleus and producing a spray of particles—a clear sign of a strong interaction. Moreover, these particles were only detected on mountaintops, not at sea level; further proof of a strong interaction with the nucleus. Once again, physicists thought they had bagged Yukawa's pion. This time they were right.

New, strange particles that left V-shaped tracks started showing up in the cloud chambers and emulsions. When a decaying particle is electrically neutral, it leaves no track in the cloud chamber (or emulsion), so the tracks of the particles produced in the decay seem to come from nowhere. In 1950, one such decay was shown to produce a proton and a pion. This was a shocker. The decaying particle, called the lambda-zero (λ^0), must be more massive than a proton and a pion combined.

By 1955, there were particles called kaons (K$^+$, K$^-$, K^0), sigma-plus (Σ^+), sigma-minus (Σ^-) and xi-minus (Ξ^-), some lighter, some much heavier than the proton. The subatomic zoo was growing larger and more confusing each year.

Thanks to Schwinger, Feynman, Dyson, and many others, QED, the theory of electrons, positrons, photons, and their interactions, was in good shape by 1950, and theorists were taking another look at the nuclear force. It occurred to more than one person that the nuclear force might behave somewhat like QED. After all, the electric force in QED was carried by photon exchange, as we have seen. In Yukawa's theory, the nuclear force was carried by pion exchange in a similar manner:

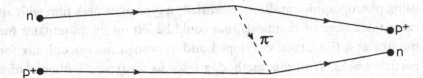

Here a neutron emits a pi-minus, turning into a proton. Then, a nearby proton absorbs the pi-minus and turns into a neutron. Thus, the nuclear force arises from pion exchange, just as the electric force in QED arises from photon exchange. To make this scheme work, theorists found they had to introduce a new concept, which they called *isospin.*

The isospin concept was introduced in analogy to an electron's spin. The electron has spin 1/2, and by the rules of quantum mechanics, it can be in one of only two states when placed in a magnetic field. The direction of the spin arrow is determined by the *right hand rule*: curl the fingers of your right hand in the direction the electron is spinning and the direction your thumb points is the direction of the spin.

Using this rule, the electron's spin is up, or +1/2, when it is in the same direction as the magnetic field and down, or −1/2, when it is opposite to the field. The down state has more energy than the up state, so it is possible for an electron with spin down to flip to spin up, emitting the excess energy as a photon.

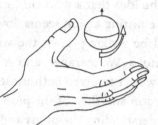

According to the isospin concept, the neutron and the proton should be thought of as two different states of the same particle, called the *nucleon*. The nucleon has isospin 1/2, which means it can exist in two different states. If the nucleon has isospin down, or −1/2, it is a neutron. If its isospin is up, or +1/2, it is a proton. By flipping its isospin, it changes from one state to the other, emitting (or absorbing) a pion in the process. In the case of the electron in the magnetic field, the photon, with spin 1, makes up for the difference between the spin-up (or +1/2) state and the spin-down (or −1/2) state. Analogously, the pion, with isospin 1, makes up for the isospin difference between the neutron and the proton. You should not think of isospin as having an actual direction in space. Unlike spin, isospin lives in a fictitious space, an "internal" direction that has nothing to do with the three-dimensional space in which we move. In the case of spin, the words *up* and *down* compare the spin direction to the magnetic field direction. In the case of isospin, the words *up* and *down* are used merely for convenience. There is no direction in physical space to compare with the isospin direction. The quantum mechanical rule is that isospin, like spin, must change in whole-number steps, so a particle with isospin 1, like the pion, can exist in three different states: isospin +1, isospin 0, or isospin −1. The +1 and −1 versions correspond to Yukawa's pi-plus and pi-minus. Isospin theory requires a third version, the neutral pi-zero, in addition.

If it sounds weird that an electron can spontaneously emit a photon, as in QED, then the idea that a neutron can spontaneously transform into a proton by emitting a pion seems downright bizarre. It is as if a Great Dane could be walking down the street and then spontaneously transform into a Weimaraner and a Siamese cat. Indeed, Yukawa's pion theory wasn't received enthusiastically at first. The discoveries, first of the muon and then of the pion itself, fired a frenzy of activity directed at understanding the pions and the nuclear forces. As Yukawa's theory began to prove its worth, the bizarre transformations it required of its particles came to seem natural and unobjectionable to physicists. Eventually, this sort of transformation became a basic feature of relativistic quantum field theory.

Building a Better BB Gun

As more and more strange and inexplicable particles were being discovered in the cosmic ray data, experimenters became impatient with waiting for something interesting to fall from the sky. How much better it would be to have a machine that produced high-energy particles! Then you would know what particle was coming in, when it was coming, what energy it had to start with, and you could be prepared to look at the results, in a way that was impossible with cosmic rays. Fortunately, a particle accelerator is easy to build. In fact, chances are you have one in your house. It's called a TV.

In order to accelerate particles, you need the following:

1. A source of particles
2. A way to speed them up
3. A way to steer them
4. A target for them to hit

In your TV, the particle source is called a cathode. It's similar to the filament in a light bulb: a little wire you heat up by passing electric current through it, which spews electrons in all directions.

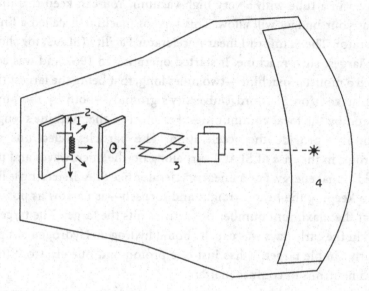

Because electrons are charged particles, all you need to accelerate them is an electric field, which is provided by connecting a voltage source to two metal plates ([2] in the diagram). After passing through a hole in the positive plate, the electrons are steered by two more pairs of plates that bend the beam in the up-down or side-to-side direction (3). The target is the phosphorescent material on the inside of the TV screen, which glows when struck by the electron beam, producing a tiny glowing dot, which forms part of the TV picture (4). The inside of the TV tube is devoid of air so that the electrons won't be scattered before hitting the screen. The particle accelerators that were being built in the 1950s (and are still being built today) are basically the same as a TV. The main difference, of course, is that the particles need to be accelerated to much higher energy in order to get anything interesting out of the collision with the target. There are two ways to do this.

One possibility is to use multiple acceleration stages, and daisy-chain them so that the particles coming out of one stage immediately enter the next stage and get accelerated some more. The whole thing is enclosed in a tube with a very high vacuum. You can keep this up as long as your budget will allow. This type of machine is called a linear accelerator. The Stanford Linear Accelerator Facility (SLAC, for short) is the largest such machine. It started operating in 1967 and was considered a monster machine—two miles long, that being the largest they could make it on Stanford University's grounds—both by the physicists and by the local communities. For most of the machine's length, microwaves provide the acceleration. The particles (electrons and positrons, in the case of SLAC) surf along on the microwaves and pick up additional energy. For a linear accelerator like SLAC, steering is limited to keeping the beam straight and focused—as narrow as possible so that the maximum number of particles hits the target. The target at SLAC in the early days was usually liquid hydrogen. Hydrogen is a particularly simple target: it has just one proton and one electron; there are no neutrons to confuse matters.

To get to even higher energy (and so probe even deeper into the proton), you might think of turning your beam around in a loop and sending it back through the accelerator, just like you might drive through the car wash twice to get your car extra-clean. This is the idea behind the cyclotron, invented in the 1920s by Ernest O. Lawrence at the University of California at Berkeley. Lawrence knew that he could use a magnetic field to get the electrons to loop around. The crucial realization came when he saw that each time the electron passed through the accelerator it would gain speed, but it would also swing out in a larger circle. Because it had farther to go, but was traveling faster, it would take the same time to go around, no matter how many times it passed through the accelerator stage. When Lawrence realized that the larger circle (of radius R) and faster speed (also proportional to R) cancelled out to keep the trip time constant, he ran around the laboratory, stopping people to tell them, "R cancels R! R cancels R!" This meant he could use a simple alternating current to provide the electric field.

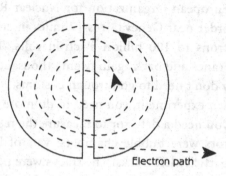

Electron path

Every time the electrons crossed the gap between the D-shaped plates, they got a small kick from the alternating electric field. By the time they returned to the gap on the other side, the electric field would have switched direction and they would receive another small kick. In this way, electrons would spiral out until they reached the outside edge of the D, at which point they would fly out in a beam. Lawrence's first machine had 4-inch Ds. With an 11-inch version, he managed in 1932 to accelerate electrons to an energy of a million electron-volts. (An *electron-volt* is the amount of energy an electron gains when accelerated by an electric potential of one volt.) At this energy, the electrons are traveling at 90 percent of the speed of light. Because of special relativity, the cancellation that got Lawrence so excited no longer works, so to go to higher energies, physicists had to do things differently.

The largest particle accelerators in existence today are called *synchrotrons*. A synchrotron is essentially a huge circular tube with magnets surrounding it to bend the electrons (or other particles) in the correct path, and with one or more accelerating stages at various points around the circle. Because the size of the circle is fixed, the magnetic field must be increased each time the particles pass through an accelerating stage, to keep them from swinging out into a larger orbit and hitting the wall of the tube. One of the most powerful machines ever built is LEP (Large Electron-Positron machine) at the European laboratory CERN (European Organization for Nuclear Research), on the Swiss-French border near Geneva. It is 17 miles in circumference and accelerates electrons to 100 billion electron-volts. They cross from Switzerland to France and back again twenty thousand times a second; fortunately, they don't need to go through customs.

To perform an experiment, you need to do more than smash particles together; you need a detector to measure the results. In the early days, the detectors were bubble chambers: vats of liquid helium in which charged particles left tracks. The tracks were photographed and the photographs carefully measured. The identity of each particle was deduced from the thickness, length, and curvature of each track. Modern detectors are much more complex and varied in design. In the most common setup, the detector surrounds the interaction region where two accelerator beams cross each other. The detector is built in multiple layers that attempt to extract all the important information about the collision. The complete onion-like structure is the size of a house.

For instance, the CDF detector at Fermilab consists of four main layers. The innermost layer is the vertex detector, constructed of thin silicon strips (somewhat like a computer chip) and designed to identify as closely as possible the location of the collision. The second layer is a large chamber crisscrossed with many exquisitely thin gold wires. This is the tracking layer. Whenever a charged particle passes near one of the wires, the wire's electrical properties change, and a computer records the changes. A strong magnetic field bends the particle's path; the curvature of the path helps identify the particle's charge and momentum. The next layer is a stack of thin lead plates interspersed with photon detectors. This layer is called the calorimeter. It is a gauntlet the particles are made to run in order to measure their energy. An electron will only penetrate a few inches, whereas a heavy particle like a proton may penetrate a foot or more. A photon, being electrically neutral, will pass through the tracking chamber without a trace, but it will deposit its energy in the calorimeter. Muons, heavier than electrons but without strong interactions, will zoom through the tracking layer, pass entirely through the calorimeter, and be trapped in the outermost layer, specially built as a muon detector. Only neutrinos run the gauntlet and escape completely, without registering in any of the layers. Their presence can be deduced from the missing energy they carry away. Many of the particles of interest don't live long enough to reach the tracking chamber. They can only be detected indirectly, by identifying the particles they decay into, and then working backward to find the energy, charge, and spin of the decaying particle.

A second method of identifying new particles looks at a whole series of scattering experiments, rather than an individual event. Experimenters tune their accelerator to a particular energy and simply record the total number of "hits," or scattering events, in which a particle from the beam is deflected or transformed into other particles. They then increase the beam energy by a small amount and again count the scattering events. The graph that results shows a bump at the energy corresponding to the mass of the particle being created in the collision, as in the following graph, which depicts the data that revealed the existence of a particle known as the J/psi.

The width of the bump, ΔE, tells the lifetime, Δt, of the particle via the Heisenberg uncertainty principle: $\Delta t = \hbar / \Delta E$. If you stand in a shower stall and sing a scale, you find that certain notes resonate much more loudly. These resonant notes occur whenever the sound wavelength fits the size and shape of the stall. Incrementally stepping the beam energy is like singing a scale. Whenever the beam energy fits the mass of a new particle (according to $E = mc^2$), there is a *resonance*: more scattering events occur. The drawback to the resonance method is that it reveals only the mass and the lifetime of the particle. To learn its other properties, spin, charge, isospin, and so forth, experimenters resort to the first method, examining individual scattering events.

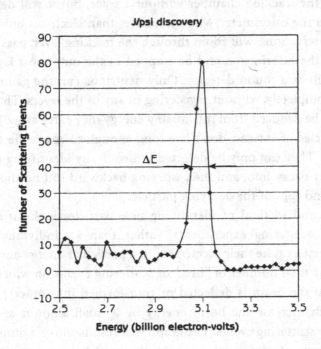

Each time a new accelerator started up, new and heavier particles were discovered. Where did these particles come from? They were created from pure energy.

Remember that, according to Einstein's $E = mc^2$, energy can be converted into matter and vice versa. The accelerator pumps some particle, say, an electron, full of kinetic energy. The electron then smashes into a target—possibly a stationary one, like the vat of liquid hydrogen in the early SLAC experiments, or else another particle, a positron, for instance, that has been accelerated so that it meets the first electron head-on. The kinetic energy of the colliding particles is released, like sparks struck off when two rocks are struck together. Now a principle comes into play that has been called the totalitarian theorem: "Anything that is not forbidden is compulsory." Applied to particle collisions, this means that anything can happen to the energy of collision, as long as it isn't forbidden by the law of conservation of energy, or conservation of momentum, or of spin, or of charge, or by some other conservation law (possibly one we haven't discovered yet). So if the colliding particles have enough energy to equal the mass, say, of a lambda-zero, then a lambda-zero *must* be produced at least some of the time, if its production is not forbidden by other conservation laws. If you had enough energy, you could, in principle, produce a Toyota the same way: pile a tremendous amount of energy into the collision of two objects, and some (infinitesimally small) fraction of the time, that energy will appear in the form of a Toyota. This is not a practical way to get a new car because, firstly, the energy needed is about that of a million nuclear bombs, and, secondly, most of the time the energy would get converted into less pleasant forms—heat, radiation, and massive destruction.

Now, suppose you are colliding electrons with positrons and you hope to produce a xi-minus, which has a charge of -1. To reach the energy you need, you are helped by the fact that the electron and the positron will likely annihilate, liberating the mass-energy of both particles as well as the kinetic energy. But there is a pesky conservation law in the way. You start out with an electron (charge -1) and a positron (charge $+1$): 0 net charge. You can't produce a xi-minus unless you also produce a positively charged particle at the same time. So we could hope, for example, to produce a xi-minus and a proton together. (Of course, we would need to begin with enough energy to add up to the total mc^2 of *both* particles.) But what if there is some other conservation law that prevents this possibility—say, conservation of isospin (if it really is conserved), or of some other type of "charge"? We can still hope to get some xi-minuses by starting with *twice* the energy we would need for one xi-minus, because we would then have enough energy to produce a xi-minus and its antiparticle, the xi-plus. Because the antiparticle has all its properties reversed, it will have the opposite isospin, and opposite values of any other possible "charges," so that the total "charge" will always be 0 for any conceivable conserved quantity. In short, we can produce any particle that exists simply by pumping enough energy into the colliding particles.

All of these possibilities can be represented by the following Feynman diagram:

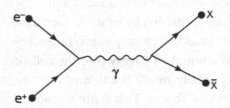

We see an electron (e^-) and a positron (e^+) approaching and colliding (left side of diagram). They annihilate, producing a photon (γ). The photon then produces a particle-antiparticle pair (X stands for the particle, $\overline{X}$ for its antiparticle). The two newly minted particles then move off in different directions.

As experimenters improved their techniques throughout the 1950s and 1960s, they discovered one new particle after another. The earlier shyness about declaring the existence of new subatomic particles wore off as the particle count ratcheted higher and higher. Physicists came to call it the *subatomic zoo*: the bewildering array of new particles with their different masses, spins, lifetimes, and decay modes. Around the turn of the century, it had seemed that three particles constituted all the matter in the universe: the proton, the neutron, and the electron. What had begun as a simple quest to explain the force that held neutrons and protons together had become a nightmare of confusion. Why were all the new particles necessary? In Rabi's phrase, who had ordered them? What did they have to do with the nuclear forces? And who could bring order to them, now that there was such an amazing variety of them?

Chapter 8

The Color of Quarks

You boil it in sawdust: you salt it in glue:
You condense it with locusts and tape:
Still keeping one principle object in view—
To preserve its symmetrical shape.
—Lewis Carroll, *The Hunting of the Snark*

By 1960, fundamental physics had gotten ugly. The beautiful simplicity of the earlier years, with every atom made of just protons, neutrons, and electrons, had yielded to the confusion of the subatomic zoo. Before 20 more years had passed, though, a new beauty would be revealed. In his novel *Perelandra*, C. S. Lewis describes a man writing by an open window who looks up in horror to see what he thinks is a hideous beetle crawling across his desk. "A second glance showed him that it was a dead leaf, moved by the breeze; and instantly the very curves and re-entrants which had made its ugliness turned into its beauties." Something much like that was to happen to elementary particle physics, the ugly riot of new particles metamorphosing into the clear signs of a deeper order in the world, a hitherto unsuspected symmetry at the heart of matter.

The new picture of the subatomic world would be the result of an intimate dialogue between experimenters and theorists. As the experimenters were bagging more specimens for the subatomic zoo throughout the 1950s, theorists were struggling to catch up. The first task was to classify the particles that were being discovered—were there groups of particles that shared some characteristics?

The first rough classification was by mass. The particles in the lightweight division, called *leptons*, were the electron, with mass about 1/2000 of the proton mass, and the neutrino, with mass of zero, as far as anyone could tell. Then there were particles with intermediate mass between the electron mass and the proton mass, like the muon and the pion. These middleweight particles were called *mesons*. The heavyweight division comprised the proton, the neutron, and all the heavier particles, which were named *baryons*.

A second classification was by spin. Particles with half-integer spin (equal to 1/2 or 1 1/2 or 2 1/2 ...) are called fermions, and they obey the Pauli exclusion principle: No two identical fermions can be in the same quantum state. We saw the importance of this property earlier; it was the property that forced electrons to arrange themselves in different energy levels and the reason the elements arranged themselves into a periodic table. Particles with integer spin (like the pion, with spin 0, or the photon, with spin 1) are called bosons. Many bosons can occupy the same state: this fact is responsible for laser light, in which all the photons of the beam act in harmony. The fermion/boson distinction reflects, to some degree, the earlier particle/field distinction. All of the particles that are normally considered matter—protons, neutrons, and electrons—are fermions. All of the intermediate force-carrying particles, such as the photon that carries the electromagnetic force and the pion that carries the nuclear force, are bosons.

The Strong, the Weak, and the Strange

By 1950, physicists had come to realize that they needed yet another classification to deal with the nuclear forces. On the one hand, there was the muon. It clearly had something to do with nuclear forces since it was produced when a nucleus suffered beta decay. However, this interaction was seemingly very weak: this was the conclusion of the famous Italian experiment discussed in the previous chapter, which showed that muons were absorbed very slowly in matter. Whatever force was involved in these muon processes must be a very weak one. On the other hand, there was the pion, which was absorbed very rapidly in matter in a manner completely consistent with Yukawa's predictions. The force carried by pions must be extremely strong, as it must hold the protons together in the nucleus while their electromagnetic repulsion is trying to tear them apart. Physicists began to speak of two different nuclear forces, called (prosaically but practically) the strong force and the weak force.

Confirmation of the strong/weak distinction came from the newly discovered subatomic particles. Some of these had extremely short lifetimes, around 10^{-24} second; comparable to the lifetime we calculated earlier for virtual pions. This was a good indication that the force involved with the decay of these particles was the same as the force involved with the pions. Other particles, the muon, for instance, had lifetimes of a few microseconds. This is very short in everyday terms, but many trillions of times longer than the lifetimes associated with the strong force. According to the Heisenberg uncertainty principle, $\Delta E \, \Delta t \geq \hbar$, a short lifetime corresponds to a large energy difference, and hence a large force; whereas a longer lifetime corresponds to a smaller energy and smaller force. So it makes sense that the strongly interacting particles should have shorter lifetimes and the weakly interacting particles should have longer lifetimes.

Some particles, like the muon, never participate in strong interactions. Other particles participate in both the strong and the weak interactions. Take the pion, for example. It certainly figures in the strong force: it's the glue that holds the nucleus together. Its lifetime, however, is around 10^{-8} second—much too long to be explained by the strong force. In fact, when it decays it produces a *muon* and an antineutrino, both of which are particles that interact only via the weak (or electromagnetic) force, never the strong force. There must be some reason that the pion, a strongly interacting particle, can only decay by the weak force. This explains both its relatively long lifetime and the particles that result from the decay. A deeper understanding of the nature of the pion had to await the development of the Standard Model.

One more classification we need to know about was invented for the strange new particles that left V-shaped tracks in the cloud chambers. Like the pion, their lifetimes were too long—around 10^{-10} second, rather than the 10^{-24} second expected from the strong interaction. But the solution found for the pion wouldn't work for the new particles. The particles they decay into are strongly interacting, so it is no good to suppose they decay via the weak interaction. How then could they be so long lived? Another strange thing about them: They always seemed to be produced in pairs, never alone. A young theorist, Murray Gell-Mann, invented a new property to help explain the peculiar properties of these new particles, which he dubbed *strangeness*. The editors of the *Physical Review* considered this name too frivolous, and Gell-Mann was forced to substitute the clunky phrase "new unstable particles," which was, according to Gell-Mann, the only phrase "sufficiently pompous" for the editors.[1] But physicists continued to call them strange particles, and eventually even the *Physical Review* had to accept that label.

Strangeness worked like this: If a strange particle decayed into a normal particle, say a proton, it was given a strangeness value of −1. A strange particle that decayed into an antiproton was given a strangeness value of +1. A strange particle that decayed into another strange particle by a slow route (10^{-10} second or so) was doubly strange, so was assigned a strangeness value of −2; whereas one that decayed by a fast route (10^{-24} second or so) was assigned the same strangeness value as the particle it decayed into. Thus, the abnormally long lifetime was explained by the fact that the strangeness value changed by 1. This ad hoc solution to the problem of the long particle lifetimes would eventually lead Gell-Mann to a crucial breakthrough.

Now that there were ways to classify the new particles, some patterns started to emerge. For instance, the leptons, the lightweight particles, are all fermions (they all have spin 1/2) and are all weakly interacting. What's more, in any interaction, the number of leptons before the interaction is equal to the number of leptons after the interaction, if we count antiparticles as having negative lepton number. For instance, in beta decay, a neutron (n) decays into a proton (p^+), an electron (e^-), and an antineutrino ($\bar{\nu}_e$).

$$n \rightarrow p^+ + e^- + \bar{\nu}_e$$

Lepton number: $0 = 0 + 1 - 1$

Physicists say that *lepton number is conserved*. Similarly, it was discovered that isospin is conserved in strong interactions, but not in weak interactions.

These patterns are not mere curiosities. From Emmy Noether's work back in the 1910s, physicists were aware that symmetry in any theory led to an invariance, that is, to a conserved quantity. Slowly, throughout the 1960s, they came to realize that the argument could also run the other way. Find the conserved quantities, and you will learn something about the symmetries of the underlying theory. It is like someone who is trying to figure out the rules of chess by looking at many different board positions, to use a favorite illustration of Richard Feynman's. They might notice that whenever one player has two bishops on the board, they are never on the same color square. This is a "conservation law" that gives a hint about what moves the bishop can make. When the observer comes up with a theory of the bishop's "interactions" (the bishop always moves on a diagonal line), the observed conservation law is seen to be a result of the interactions. As in chess, it is not necessarily easy to guess the particle interactions from the symmetries. (You can probably think up other theories about the bishop's motion that would still keep the two bishops on opposite-colored squares.) In 1959, Murray Gell-Mann began to tackle the problem of the strong interactions from the point of view of symmetry.

A New Periodic Table

As a young man, Gell-Mann was every bit the precocious genius that Schwinger had been, but even more impressive was the breadth of his interests and knowledge. He was as likely to hold forth on linguistics, history, archaeology, or exotic birds as on physics. He had a tremendous talent for languages and was famous for correcting people on the pronunciation of their own names. For all his intellectual arrogance, he was often curiously reluctant to publish his ideas. He said once that a theoretical physicist should be judged by the number of right ideas he published minus twice the number of wrong ones. His paper introducing the idea of strangeness was circulated to colleagues, but never published. His was unquestionably one of the most fertile minds of his generation and his enthusiasm for physics was unbounded. He once told an interviewer, "If a child grows up to be a scientist, he finds that he is paid to play all day the most exciting game ever devised by mankind."[2]

As the count of fundamental particles blossomed from three in 1932 to 16 in 1958, physicists found themselves in the same position as Mendeleev in 1869, who had had 62 chemical elements and no good classification scheme. The scheme that Mendeleev developed, known to us as the periodic table of the elements, not only classified the elements according to their chemical properties but allowed Mendeleev to predict the existence of as yet undiscovered elements by looking at the gaps in his table. As we have seen, the whole structure of the periodic table would ultimately be explained in terms of the three particles that are the building blocks of the elements, the proton, the neutron, and the electron. What physicists needed was a new periodic table for the fundamental particles. In 1961, Murray Gell-Mann gave them one.

Gell-Mann's initial step seems simple enough in retrospect: He plotted the known particles on a diagram with a particle's isospin on one axis, and the newly invented property, strangeness, on the other. For the eight lightest mesons, the result came out like this:

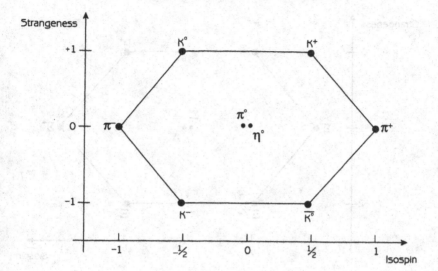

The result: a perfect hexagon. There are two particles at the center, both with isospin 0 and strangeness 0, so these particles form an octuplet, leading Gell-Mann to dub his scheme the *Eightfold Way*, a light-hearted reference to the Buddhist teaching:

> Now this, O monks, is the noble truth that leads to the cessation of pain, this is the noble Eightfold Way: namely, right news, right intention, right speech, right action, right living, right effort, right mindfulness, right concentration.[3]

To get his scheme to work, Gell-Mann had to assume that the K^0 had strangeness +1, but its antiparticle, the $\overline{K}^0$, had strangeness −1, in other words, they were two distinct particles. The referees reviewing his paper for publication objected: Neutral mesons like K^0 were supposed to be their own antiparticle. "It's all right," Gell-Mann replied, "they can be like that."[4] As it turned out, he was right; they *could* be like that.

What about the other particles? The lightest baryons fell out like this:

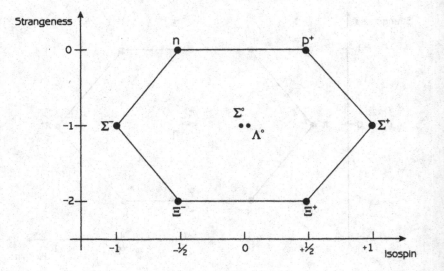

Another octuplet. The Eightfold Way was working. It was all very well to see the particles falling into pretty patterns on the strangeness-isospin diagram, but what did it mean? Was there some mathematical relationship that determined these particle multiplets? And, did this new periodic table hint at some substructure for the particles in the same way that the periodic table of the elements was explained, many years later, by the discovery of atomic structure?

Gell-Mann tackled the mathematical question first. Working by analogy to the well-understood example of spin, he tried to build up his octuplets, but without success. A chance conversation with a mathematician colleague led him to an obscure bit of nineteenth century mathematics called Lie groups, after the Norwegian mathematician Sophus Lie (pronounced "Lee"). Here he found what he wanted: a Lie group that had an octuplet representation. This group was called SU(3). The mathematics of group theory (which we will skip over) guaranteed that the masses of different particles in the octuplets had to be related in certain ways. Gell-Mann checked these relationships and found them to be satisfied for the octuplets he had constructed. These mass rules were to play a crucial role in a dramatic discovery.

One of the triumphs of Mendeleev's periodic table was his ability to predict new elements from the gaps in his table. At a conference of particle physicists in Geneva in 1962, Gell-Mann got the opportunity to predict a new particle using the Eightfold Way. The Lie group of the Eightfold Way, SU(3) allows other representations besides the octet. There is a representation with 10 particles (decuplet), one with 27 particles, and so on. Gell-Mann and his colleagues had been worrying about what to do with a set of four particles, called the delta particles($\Delta^-, \Delta^0, \Delta^+, \Delta^{++}$), that had been known since 1952. They wouldn't fit in an octet, but they would fit either the decuplet or the 27-particle multiplet. At the 1962 conference, two new particles were reported that allowed Gell-Mann to create this diagram:

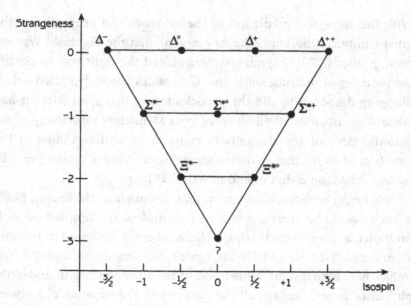

This was the expected form for the decuplet, and with the other newly discovered particles, only the bottom corner of the triangle was missing. On the spot, Gell-Mann went to the blackboard and predicted a new particle, which he called the omega-minus (Ω^-), and described what its properties should be: charge -1, spin $3/2$, mass about 1670 million electron-volts. By 1963 it had been detected, with the predicted properties, at an accelerator in Brookhaven, NY. The Eightfold Way was a success, and Gell-Mann was the new Mendeleev. He had to share the glory, however, which he did readily. An Israeli politician, Yuval Ne'eman, who studied physics when he could get away from his job of defense attaché, had come up with the same prediction at the same 1962 conference. The two had never met before the conference, but, according to Ne'eman, "from then on, they became close friends."[5]

Three Quarks for Muster Mark

With the successful prediction of the existence and properties of the omega-minus, it became clear to everyone that the Eightfold Way, or, more properly, SU(3) symmetry, was indeed the right way to classify the particles of the subatomic zoo. Gell-Mann hardly hesitated before plunging deeper. Why did the particles follow that symmetry? It had taken scientists more than 50 years from Mendeleev's discovery of the periodic table of the elements to come to an understanding of the structure of atoms that answered the question: Why is there a periodic table? Gell-Mann didn't intend to wait that long.

He began with the observation that, although all the known heavy particles could be arranged in SU(3) multiplets, the simplest possible multiplet, a three-particle triplet, was nowhere to be found in the subatomic zoo. Why should Nature ignore the simplest solution? It was unlike her. Perhaps the triplet was there all along, right under the physicists' noses. Perhaps all the particles in the subatomic zoo were built out of these three fundamental particles. He decided to call the quirky little things *quarks*, and was delighted to find support for the name in a passage from James Joyce's *Finnegan's Wake*:

> Three quarks for Muster Mark.
> Sure he hasn't got much of a bark
> And sure as any he has it's all beside the mark.

To make the scheme work, though, Gell-Mann had to make a radical assumption. Every particle ever detected had an electric charge that was a whole multiple of the electron charge. Reluctantly, Gell-Mann concluded that his quarks had to have electric charges that were *fractions* of the electron charge. The quarks, which he named up, down (in analogy with spin-up and spin-down), and strange (because it was a constituent of the strange particles) had charges of +2/3, −1/3, −1/3, and isospins of +1/2, −1/2, 0, respectively.

Gell-Mann was well aware of the objections his colleagues would raise: Why have we not seen these particles? Why have we never seen *any* fractionally charged particle? He was uncharacteristically timid in the paper he wrote proposing this model, giving just the barest outline, and ending the paper with the enigmatic phrase: "A search for stable quarks...would help to reassure us of the non-existence of real quarks."[6] He seemed to be saying that quarks were just a mathematical abstraction that allowed the properties of the known particles to be explained. His reluctance to say clearly whether or not quarks were real physical entities was summarized by a colleague like this: "If quarks are not found, remember I never said they would be, if they are found, remember I thought of them first."[7]

Actually, someone else had thought of them independently. George Zweig had written an 80-page paper that worked out many of the consequences of the quark model (he called them aces), which was not published until years later. Zweig, like Gell-Mann, was cautious about claiming these bizarre particles actually existed, but he was bold enough to end his paper on an optimistic note: "There is also the outside chance that the model is a closer approximation to nature than we may think, and that fractionally charged aces abound within us."

Today, both are given credit for the model, but it is Gell-Mann's names that are used.

The difficulty in deciding whether quarks were real lay in the very idea of a fundamental particle. The term "fundamental" was being used in two different senses. One meaning is *basic constituent*: a particle that is a building block for more complex objects. A second meaning is *particle that cannot be broken apart into smaller particles*. It had long been accepted that protons, neutrons, and electrons were the basic constituents of atoms, a view that was confirmed by experiments that split atoms apart, yielding the particles themselves. It was natural to think that, if protons and neutrons were made of still smaller constituents, it should be possible to split the proton and observe a free, fractionally charged quark. Since no such particle had ever been detected in all the years of accelerator experiments, Gell-Mann's reluctance to predict the existence of quarks as physical particles is understandable. As we will see, it would take a better understanding of the interactions between quarks to explain why quarks could be basic constituents of protons and neutrons but nonetheless be trapped forever inside them, unable to be knocked out as free particles regardless of how energetically the protons and neutrons were slammed into each other.

With the quark model in hand, the entire subatomic zoo suddenly started to make sense. First off, all of the baryons (the heavyweight particles like the proton and neutron) are composed of three quarks. For instance, the proton is made of two up quarks and one down quark. This gives the correct electric charge, since $2/3 + 2/3 - 1/3 = +1$, as well as the correct isospin: $1/2 + 1/2 - 1/2 = 1/2$. The neutron has one up quark and two down quarks: Its charge is $2/3 - 1/3 - 1/3 = 0$, and its isospin $1/2 - 1/2 - 1/2 = -1/2$, as it must be. Strange particles contain a strange or antistrange quark, and doubly strange particles contain two, and so on. The quark structure of particles is discussed in more detail in Appendix A.

The mesons (the middleweight division) are built from one quark and one antiquark. The three pions, for example, are made of various combinations of up and down quarks and their antiquarks.

The leptons (the lightweight division) are the particles that are not built from quarks. In this new understanding of leptons, the muon had to be reclassified. Even though its mass put it in the middleweight division, it didn't participate in the strong interaction. In the Gell-Mann–Zweig quark model, all quarks participate in the strong interaction. The muon therefore could not be built of quarks. Even though it is 200 times heavier than the electron, it must be considered a fundamental particle with no structure. In this new understanding, the muon is a lepton, not a meson.

The quark model not only explained the observed properties of the particles in the subatomic zoo, it also gave a clear picture of their interactions. Physicists had known since the 1930s that the pion held neutrons and protons together in the atomic nucleus in much the same way that photons cause the attraction of oppositely charged particles. This was the premise of Yukawa's theory of the strong interactions. Recall how pion exchange looks in terms of Feynman diagrams. The proton emits a pi-plus and changes into a neutron. The neutron then absorbs the pion and changes into a proton.

The quark model gives a deeper understanding of the pion-exchange process.

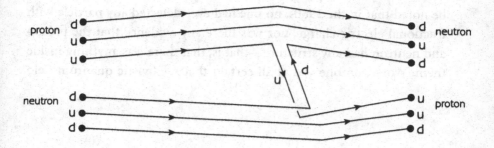

Here, we see a proton (duu) and a neutron (dud) coming in from the left. (In this notation, which we will continue to use, d is a down quark, u an up quark, and an antiparticle is represented with a bar over its letter or symbol.) At the upper kink in the diagram, a virtual $d\bar{d}$ pair is created, in the same way that virtual electron-positron pairs are created from empty space. (There must, of course, be a photon or some other particle, not shown in the figure, that provides the energy to create this quark-antiquark pair. For now, we'll set aside the question of what sort of particle this might be.) The $u\bar{d}$ combination is emitted, leaving a neutron (dud) on the right. The pion ($u\bar{d}$) is then absorbed by the original neutron. This, we now see, is nothing more than $d\bar{d}$ annihilation. What's left (uud) is a proton. The result is the same as in Yukawa's theory, but it is now understood in terms of the production and annihilation of quark-antiquark pairs.

It's all very well that the particles and their (strong) interactions can be understood in terms of the quark model, but what about the question that Gell-Mann was so ambivalent about: Are quarks real? In favor of the affirmative, Gell-Mann and Zweig could point out that all of the known strongly interacting particles could be explained in terms of three fundamental quarks. We now know that there are actually six quarks, called up, down, strange, charm, top, and bottom, which are known as the quark *flavors*. We can still say today that all known strongly interacting particles can be built out of these six quarks and, of course, their antiquarks. On the side of caution, however, it should be noted that in the 1960s, no one had ever detected any particle with fractional electric charge, nor was there any evidence that the proton and neutron had any structure—that is, that there was anything inside them. Worse, no one was at all certain that relativistic quantum field theory was the right approach for understanding the strong interaction. There was a plethora of alternatives: the S-matrix, Regge poles, the SU(6) group. No one, not even the quarks' inventors, would have bet money that they were real, physical particles.

Evidence that quarks actually existed began showing up in experiments at Stanford University in 1968. The SLAC machine accelerated a beam of electrons and fired it into a vat of liquid hydrogen, a procedure considered uninteresting by physicists on the East coast, who had already begun colliding protons with each other head-on. The proton-proton collisions were more energetic, but they were messy; it was the cleaner electron-proton collisions at SLAC that would be the first to reveal the structure of the proton.

Think back to the example of the BBs fired at the Nerf ball. The electrons (the BBs) were deflected by the protons (the Nerf ball) and the pattern of the deflection told experimenters about the internal structure of the proton. The interpretation of the results can be quite tricky, however, especially when there is more than one object inside the target particle. There is no way to hold the quarks in place while firing many electrons to map out the locations of the quarks. Each scattering event involves a potentially different arrangement of quarks inside the proton. The most that experimenters could say was that the size of the proton was about 1/1000 the size of a typical atom, and that the proton had no hard core; there was no "pit" inside the "peach." At the tremendously high energies (for the time) reached by SLAC, there was another kind of scattering that could occur. The energy that the electron carried could be converted into other particles, a process known as inelastic scattering. The two scattering possibilities, elastic and inelastic, can be diagrammed like this:

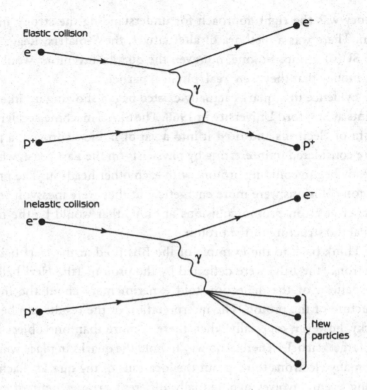

Surprisingly, it was the messier inelastic collisions that were shown to have a simple property, called *scaling*, that provided the evidence that protons were really made of point-like constituents: the quarks. In any scattering process, the probability of the incoming electron scattering in a particular direction depends on two quantities: the energy of the incoming electron and the amount of energy the electron loses. Very roughly, scaling refers to the fact that at high enough energy, the scattering probability no longer depends on the energy loss. This was a surprise to the experimenters who discovered it. If the proton had some complicated internal structure, you would expect the scattering probability to get more complicated as you go to higher energy and probe the interior more deeply. Instead, things got simpler. Why?

Although he didn't invent the idea, Richard Feynman was the person who drew attention to scaling in inelastic collisions, and he was the one who explained the phenomenon in a simple and convincing way. Feynman didn't want to rely on the quark model; he didn't think that there was sufficient evidence for quarks in 1968 to take them as proven. He decided to start from the assumption that the proton was made of some collection of point particles, while remaining uncommitted about their numbers and interactions, and see what predictions would follow from these minimal assumptions. He didn't want to confuse his simple model with the more sophisticated quark theory, so he called his hypothetical point particles partons, as they were the parts of the proton. Feynman found that the parton model was enough to derive scaling. To put it another way, the scaling that experimenters were finding in their data was evidence that the proton contained point-like particles of some kind. Did this mean scaling was evidence of quarks? Feynman thought not, but Gell-Mann vehemently disagreed. "The whole idea of saying they weren't quarks and anti-quarks but some new thing called 'put-ons' seemed to me an insult to the whole idea that we had developed," Gell-Mann said later. "It made me furious, all this talk about put-ons."[9]

Physicists did not immediately rush to embrace the quark model when evidence of scaling showed up in the SLAC experiments. A great deal of effort had been invested in alternate approaches, and those who had invested it were understandably reluctant to throw it all away for something as iffy as field theory. The quark model was not without its problems, either. For one thing, the scaling idea depended on the assumption that the quarks were weakly bound, nearly uninteracting. But in that case, it should be easy to knock a quark out of the proton and detect it on its own. In spite of much intense searching, nothing resembling a free quark had ever been seen. How could quarks be nearly uninteracting and yet be confined inside the proton?

Another problem with the quark model can be seen in the famous omega-minus (Ω^-) particle that Gell-Mann predicted using the Eightfold Way. In the quark model, it should be made of three strange quarks, each with a charge of −1/3 and spin of 1/2. The omega-minus has spin 3/2, so the spins of the three quarks must be aligned (that is, all must be spin up or all must be spin down). To understand why the very existence of such a particle was puzzling, recall the Pauli exclusion principle: You can never have two identical fermions in the same quantum state. The exclusion principle was the reason each energy state of an atom can only accommodate two electrons: the electrons are spin 1/2 fermions, and so have only two possible states, spin up or spin down. Now, strange quarks are also spin 1/2 fermions, so they also can only be spin-up or spin-down. But that means that the quark picture of the omega-minus is impossible. Three identical quarks can never have their spins in the same direction.

Finally, and perhaps most importantly, the quark model said nothing about the forces that held the quarks together in the proton and the neutron. The strong force that bound protons and neutrons in the nucleus was now understood to be caused by an exchange of pions, which is the same thing as exchanging a quark-antiquark pair, as we have seen. But there had to be yet another force that operated inside the proton that kept its quarks from flying apart.

The solution to all three puzzles came in 1972 from an old hand applying an old idea with a new twist. The old hand was Murray Gell-Mann, working together with his colleagues Harold Fritsch and William Bardeen. The old idea was the same SU(3) group that had explained the Eightfold Way classification of the subatomic particles. Gell-Mann and colleagues, however, this time used it in a completely new way. The basic idea is this: Each flavor of quark, up, down, strange, and so on, comes in three colors, which were given the names red, green, and blue, after the primary colors of light. So there is a red up quark, a blue up quark, and a green up quark, and so on for each quark flavor. This is not a claim that the quarks would actually look those colors if you could see them—rather, the quarks have an additional, and heretofore undiscovered, property, which is given the arbitrary name color. It could just as well have been called anything else that comes in threes. Gender (he, she, it), for instance, or porridge (hot, cold, just right). Color, however, turns out to be a particularly apt metaphor.

Straightaway we can see how the new quantity, color, solves the omega-minus puzzle: simply require that each strange quark in the omega-minus have a different color. The Pauli exclusion principle requires that the three quarks all be in different quantum states. The new property, color, allows us to give them different color states, so the fact that they all have the same spin state is no longer a problem.

Just as the three primary colors of light combine to make white light, the three quark colors combine to make "colorless" particles. We can also make an analogy with electric charge. A hydrogen atom is made of a negatively charged electron and a positively charged proton. From a distance, the positive and negative charges appear to cancel, and the atom looks neutral, or uncharged. It is only when you probe deeply into the atom—for instance, by shooting electrons at it as Rutherford did—that you discover the charged particles hidden inside. In the same way, a proton might be made up of a red up quark, a blue up quark, and a green down quark. From a distance, the three color charges cancel each other, and the proton appears "white," or color-neutral.

The mesons work a little differently. Each meson is made of a quark and an antiquark, and the antiquarks come in the colors antired, antiblue, and antigreen. So a pion, the pi-plus, for instance, might be composed of a red up quark and an antired antidown quark.

According to Gell-Mann & Co., only color-neutral combinations appear in free particles. This was why color had never been detected. An electrically neutral particle doesn't produce or respond to an electric field. In the same way, a colorless particle doesn't produce any external signature. Since all free particles are, by fiat, colorless, no color force has ever been observed. This may seem like a cheat: introduce a new quantity that solves the immediate problem but is otherwise unobservable. However, in the form introduced by Gell-Mann, Fritsch, and Bardeen, it was much more than that. The color theory also describes the force holding the quarks together inside the proton. The theory predicted that this force, the glue that holds the proton together, would be carried by a set of new, massless particles, in the same way that the electromagnetic force is carried by photons. With Gell-Mann's typical flair for names, he dubbed these new particles *gluons*. The whole theory of interacting quarks and gluons he called *quantum chromodynamics* or QCD.

The discovery of scaling, together with Feynman's parton explanation of it, gave theorists and experimenters alike reason to think that the proton was built of smaller point-like particles. About the whole apparatus of color theory and QCD, though, they remained skeptical. Particularly troubling was the claim that these quarks were confined forever inside the protons and neutrons. How could anyone believe in particles that could never be observed on their own? Everything changed in the November Revolution of 1974.

At the time, three quark flavors were all that were needed to explain the properties of the known particles. Protons and neutrons were composed of up and down quarks, and strange particles required a third flavor called, logically enough, the strange quark. A few theorists played around with the idea that quarks should be organized into two-quark families. The lightest quark family consisted of the up quark and the down quark. To complete the second family, a fourth quark was needed to go along with the strange quark. They whimsically termed this hypothetical new quark charm.

The charm quark seemed to solve some difficulties of the weak interactions. It all seemed very tenuous, however. *If* you believed in quarks, and *if* you believed they could be described using color theory, and *if* the weak interactions could also be described using color theory, and *if*, on top of all that, you bought into the idea of a fourth quark that was otherwise completely unnecessary, then a few little theoretical puzzles went away. Then, in September 1974, Samuel Ting and his experimental group at Brookhaven National Accelerator Laboratory on Long Island found a bump in their data that indicated the existence of a new particle, which Ting named the J. A meticulous experimenter, Ting had been taking data using very small steps in energy, and so had caught sight of a narrow resonance that other, more powerful accelerators had missed. But now Ting's natural caution tripped him up. He passed up several chances to announce the discovery, wanting first to eliminate any possibility of error. As he waited, word leaked out about the bump in the data at an energy of 3.1 billion electron-volts. A team at Stanford Linear Accelerator Center (SLAC) found the same bump in early November, assigning it the Greek letter psi. When Ting heard that the SLAC group was about to announce its discovery, he scrambled his group to make its own announcement. The two groups' papers were published simultaneously in the December 2 issue of *Physical Review Letters*, and the particle is now known as the J/psi, pronounced "Jay-sigh."

The importance of the J/psi discovery lay in its theoretical interpretation: It was charmonium, a charm quark tightly bound to an anticharm quark. The evidence for this interpretation lay first of all in the narrowness of the bump in the energy spectrum (see the graph shown in the previous chapter, in the section "Building a Better BB Gun"), which implied a lifetime about a thousand times longer than expected for a strongly interacting particle. Secondly, charmonium should theoretically exist in several different configurations. Experimentally, this meant that the big bump at 3.1 billion electron-volts should consist of several separate bumps. When experimenters confirmed the existence of some of these individual resonances, the charmonium model quickly became accepted as the correct explanation of the J/psi.

After the November Revolution, interest in the quark model exploded. Experimenters competed for the glory of discovering particles that combined a charm quark with an up, down, or strange quark, as theorists frantically set about calculating their expected properties. In 1977, another narrow resonance suggested the existence of a fifth quark. This in turn implied a third quark family. Some pushed for the fifth and sixth quarks to be called "beauty" and "truth," but eventually the names "bottom" and "top" won out. In experiment after experiment, the quark model proved its worth. By the time the discovery of the top quark was announced in 1994, there was no longer any doubt that quarks were real.

The World According to Quarks

QCD tells us that each flavor of quark comes in three colors. We can depict this as a color triangle, here drawn for the up quark:
The SU(3) symmetry of QCD says that we can rotate the triangle, and

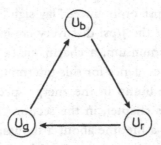

the theory will remain unchanged. In other words, if we could reach inside every proton and neutron in the universe and instantaneously replace every red up quark with a green up quark, every green up quark with a blue up quark, and every blue up quark with a red up quark, and simultaneously do the same with all the other quark flavors, there would be no way to tell that we had done so. The universe would continue on exactly as before the change.

According to Noether's theorem, this kind of symmetry guarantees that color is a conserved quantity. That means that in any interaction, the amount of color going into the reaction equals the amount coming out. Now, QCD tells us that a quark can emit a gluon and change its color:

Here the double wavy line represents the gluon. But color is not

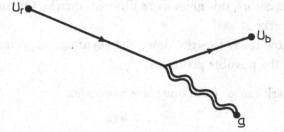

conserved in this diagram unless the gluon carries color. In fact, the gluon has to carry two kinds of color: It must carry red (so that the outgoing colors include the red of the incoming quark) and antiblue (to cancel out the blue of the outgoing quark). Let's add color to the Quark and gluon lines, using a thick line for red, a medium for green, and a thin for blue. The antiblue in the gluon double line is represented by a backward arrow:

Now we can see that color is conserved: red comes in, and (red + blue

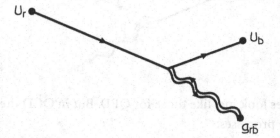

+ antiblue = red) comes out.

This is a major difference between QED and QCD. In QED, the photon that mediates the interaction between electrically charged particles is not itself electrically charged. In contrast, the gluon of QCD, which mediates the interaction between the colorful quarks, itself carries color. As a result, gluons can interact with each other, something photons can't do.

QCD doesn't tell us anything about the possible flavors of quarks;

we simply have to go looking for them among the particles of the sub-atomic zoo. We now know that there are six quark flavors. Together with the three colors, this gives us 18 different quarks. Of course, there are 18 antiquarks as well.

We are now ready to write down the Feynman diagrams for QCD that show all the possible processes.

- A quark can go from one place to another.

- A gluon can go from one place to another.

- A quark can absorb or emit a gluon.

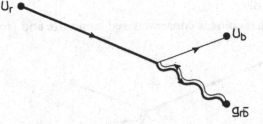

So far, the rules look just like those for QED. But in QCD there are two more possible processes.

- A gluon can split into two gluons.

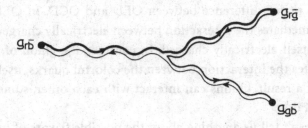

■ Two gluons can collide and form two new gluons.

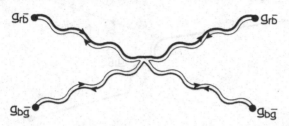

These additional interactions fundamentally change the behavior of the theory and ultimately explain why we never see free quarks in nature.

Now two of the three major chunks of the Standard Model are in place. The first chunk is QED, the relativistic quantum field theory that describes the interactions of charged particles and photons. The second chunk is QCD, the relativistic quantum field theory of quarks and gluons. The only thing missing from our Theory of Almost Everything is a description of the weak interactions—the subject of the next chapter.

For now, let's see where our description of the world has brought us. Everything around us is composed of atoms, which consist of a small nucleus of protons and neutrons surrounded by a cloud of electrons. The electrons are bound to the nucleus via photon exchange, as described by QED. In the nucleus, protons and neutrons are bound together via pion exchange, which we now know is actually the exchange of a quark–antiquark pair. Each proton or neutron is composed of three quarks, which are bound together by gluon exchange, in other words, the color force. Inside the proton there is a whirlwind of activity: gluons being exchanged, virtual quark–antiquark pairs being created and annihilating. The quarks themselves come in six flavors and three colors, for 18 versions, or 36 versions if you count the antiquarks as well.

We are now ready to tackle the remaining puzzle of the quark model: why do quarks behave as if they are weakly bound in scattering experiments, yet they can never be knocked free? Here is the basic gluon-exchange interaction between two quarks:

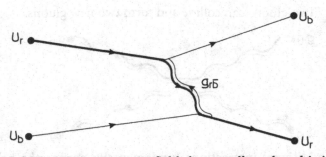

The rules of relativistic quantum field theory tell us that this interaction will be modified by the creation of virtual quark-antiquark pairs, virtual gluons, and so forth. Remember the totalitarian theorem: Anything that is not forbidden is compulsory. For QED, we discovered that the cloud of virtual particles has the effect of screening the electric field. This means that as we go to higher energy and penetrate deeper into the virtual cloud the electric force increases more than we would have expected had we not known about the virtual particles. For QCD, we find exactly the opposite effect. That is, as we go to higher energy and shorter distances, the strength of the color force decreases. (The details are worked out in Appendix B.) Conversely, at longer distances, the strength of the color force increases. It's like putting two fingers inside a rubber band to stretch it. The farther apart you pull your fingers, the larger the force. But if you move your fingers together, the rubber band goes slack. This property of QCD is called *asymptotic freedom*, meaning that at the extremes of large energy the quarks behave like free, non-interacting particles. The discovery of asymptotic freedom earned David Gross, David Politzer, and Frank Wilczek the 2004 Nobel Prize in Physics.

At last we can understand why quarks can act like free particles in high-energy scattering experiments, yet remain bound in the proton and neutron. At the high energies of the scattering experiments, the interaction strength is less, and the quarks behave like free particles. On their own inside a proton, the quarks have much lower energy, so the interaction strength is higher, and the quarks remain bound. Roughly speaking, the strong force is like stretching a spring: When the quarks are close to each other, the spring is unstretched, and the force is small. As we try to pull the quarks apart, the force gets larger and larger, making it impossible to pull a quark out of a proton far enough to detect it as a free, fractionally charged particle. The strength of the QCD interaction has now been measured over a wide range of interaction energies, and the results are in striking agreement with the theoretical prediction.

How does this explain why we never see a lone quark? Let's start by picturing a pion, say, as a quark and an antiquark connected by a tube of color, like two marbles connected by a rubber band (top, in the diagram below). Just as two electrically charged particles have a tube of electric field connecting them, the two quarks have a tube of color field. Just as an electric field can be thought of as made up of virtual photons, the tube of color can be thought of as a stream of virtual gluons. As we pull the quarks apart, we have to add more and more energy until the tube eventually snaps (middle). The breaking point occurs when the added energy becomes sufficient to create a quark-antiquark pair, so that instead of two free quarks, we end up with two new mesons (bottom).

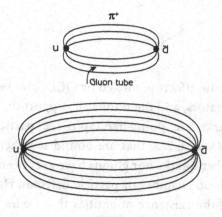

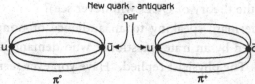

Physicists call this *color confinement*: Particles with color can only occur in color-neutral combinations. Free particles with color are never seen.

The process shown here is not the only one possible. Usually, there is so much energy in the gluon tube when it snaps that numerous particle-antiparticle pairs are produced, resulting in particle jets in the directions that the original quarks were moving. Jets like these have actually been observed at several accelerators, providing dramatic (but indirect) evidence for the existence of the elusive quarks.

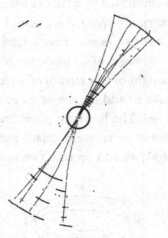

Other, more subtle effects predicted by QCD have been tested at many different accelerators, and the results all support the picture that QCD describes: Quarks are point-like, spin-1/2 particles, with electric charges of ± 1/3 or ± 2/3, that are bound together by the gluons of QCD. Still, neither quarks nor gluons have ever been directly observed as tracks in a cloud chamber or particle detector. How then can we be so confident of the existence of entities that we have never seen (and according to the theory of QCD, will never see)?

Leon Lederman, speaking to an audience of nonphysicists, was once confronted by an irate questioner who demanded an answer to that question. The physicist replied, "Have you ever seen the Pope?"

"No," was the reply.

"Then how do you know he exists?"

"Well, I've seen him on TV," the interlocutor responded.[10]

Pause for a moment and think what this answer implies. A device (the TV camera) admits light and converts it into a code involving tiny magnetic areas on a piece of videotape. Some time later, a TV station takes that tape and converts it into a stream of electromagnetic waves or a cable signal, which carries the information to your TV. Your TV then turns this code into a series of minute variations in the sweep of an electron beam, resulting in an image on the phosphorescent screen of the TV. By comparison, the indirect evidence for quarks and gluons is rather straightforward. The point is this: If each step in the chain of evidence is well understood, then indirect evidence carries nearly as much weight as direct evidence. If we can believe there is a Pope because we've seen him indirectly on TV, then we can believe in quarks because we've "seen" them indirectly through particle jets, through scaling, and through their amazing capability to explain the properties of strongly interacting particles.

Ignorance Is Strength

The history of theories of the strong force reveals how important the concept of symmetry has become in physics. The early isospin theory was an expression of the (inexact) symmetry between neutrons and protons. The Eightfold Way extended that symmetry in order to include the different types of heavy particles that were turning up in accelerator experiments. In the triumphant version of the strong force theory, QCD, symmetry and force are intimately and inextricably intertwined.

Call it the principle of *ignorance is strength*. Color SU(3) symmetry declares that there is no way to tell the difference between, say, a red up quark and a blue up quark. Now, suppose I could isolate a single quark and store it in my lab in some kind of box. I then declare that this is a color standard; this quark is the definition of the color red. (The particle property, not, of course, the visible color.) In the same way that anyone can determine if they have a kilogram of mass by comparing it to the standard kilogram in Paris, anyone who wants to know the color of a quark has only to compare it to my standard quark. Let the two quarks interact, and it will be possible, in principle at least, to determine the color of the second quark.

Now, suppose that experimenters at another laboratory on a different continent isolate their own quark and declare *it* the definition of the color red. They can perform any experiments they like using their definition of red and obtain results that are completely consistent with the results coming out of my laboratory, whether or not their "red" is the same as my "red." In order to find out if we are talking about the same red, I only need to pack my standard quark and fly to their laboratory to compare with their standard quark.

In other words, since red, blue, and green quarks are completely equivalent, the term red can have different meanings at different locations. It is only when the two quarks are at the *same point in space* that they can interact and so be compared. There is nothing to prevent us, then, from defining red differently at *every point in space*. We only need to keep track of how the definition of red changes from point to point. The mathematical function that keeps track of the changes in the definition is something we have encountered before—it is a *field*. In fact, for QCD, it is the gluon field. Gluons, therefore, are a manifestation of our inability to distinguish a red quark from a blue quark. That is, the color *force* is a consequence of the color *symmetry*. Ignorance is strength.

The ignorance is strength principle is the basis of all the forces in the Standard Model. Theories that work this way are called Yang-Mills theories or gauge theories. At the root of a Yang-Mills theory is a symmetry. If we make one assumption—that the symmetry transformation can be applied differently at different points in space—then we find mathematically that the symmetry guarantees the existence of a quantum field that mediates a force. In QCD, the symmetry is color symmetry, and the quantum field that arises from this symmetry is the gluon field that binds the quarks together into particles. Ultimately, color symmetry is the reason that matter as we know it can exist.

QED is another example of a Yang-Mills theory. The terms positive and negative for electric charge are just as arbitrary as the terms red, blue, and green for the color force. Our inability to define the charge in any absolute sense is the ignorance that gives rise to a quantum field: the photon field. This symmetry (called U(1) symmetry) guarantees that electric charge will be conserved—Noether's theorem at work. But it does even more; it guarantees the existence of the photon, the quantum of light. This symmetry is the reason light can travel across a billion light-years of empty space, bringing us news of distant galaxies.

The crucial idea is that QED and QCD are the same type of theory, that both the electromagnetic force and the strong force are related to the symmetries of the fundamental particles involved (the electrons and quarks). The reason light and electromagnetic forces exist is the underlying U(1) symmetry of charged particles. The reason quarks stick together in protons and neutrons is the underlying color symmetry. As we will see shortly, the Standard Model is built on the same central idea. In the Standard Model, symmetry is force and force is symmetry.

Chapter 9

The Weakest Link

And so some day
The mighty ramparts of the mighty universe
Ringed round with hostile force
Will yield and face decay and come crumbling to ruins.
—Lucretius, *De Rerum Natura*

At this point, it might seem like the story is over, the Theory of Almost Everything complete:

- There are six flavors of quarks, each coming in three colors.

- Ordinary matter contains only the two lightest quark flavors, up and down. Two ups and a down form a proton, two downs and an up form a neutron. Gluons bind the quarks together.

- Neutrons and protons are themselves bound together in atomic nuclei by pion exchange, which is to say exchange of quark-antiquark pairs.

- The electromagnetic force, in other words, photon exchange (or QED), holds electrons in atomic orbitals.

- The Pauli exclusion principle determines how the electrons may be arranged in the orbitals, which in turn determines the atom's chemical properties and its place in the periodic table.

■ The chemical properties of the atom determine all its interactions with other atoms. These interactions, together with the atomic masses, determine all the everyday properties of objects: density, texture, inflammability, frangibility, electrical conductivity, even color.

Apparently, this is everything we need to explain all everyday phenomena through the combination of QCD and QED. There is, however, one more piece to the puzzle. The missing piece has nothing to do with the strength of steel or the density of gold, but it is crucial for life on earth even to exist.

The story of the weak nuclear force (or weak force, for short) begins in the late nineteenth century with the discovery of radioactivity. One type of radioactivity resulted in the emission of beta rays, later identified as electrons. Where did the electrons come from? These were not the electrons in orbitals around the atom. Instead, they were produced by the nucleus, which changed its identity and became a different element in the process. In 1933, the caustic Italian physicist Enrico Fermi shocked the physics community by proposing that beta radiation occurred when a neutron decayed into an electron, a proton, and a new, never-detected particle that we now call an antineutrino. The shock was two-fold: First, predicting particles on purely theoretical grounds was still a recent and somewhat disreputable innovation (this was long before Gell-Mann, of course), and second, it was almost inconceivable that a neutron, at that time considered a fundamental particle, could decay.

To understand the need for the antineutrino, think about what would happen without it: The rest energy of the proton and electron add up to less than the rest energy of the neutron. This means that when the neutron decays there is some energy left over, which must go into the kinetic energy of the motion of the proton and electron. Taken together, conservation of energy and conservation of momentum completely fix the electron's energy: All beta decays should produce electrons with the same energy. This is not what is found experimentally, however. The electrons from beta decay had a wide range of energies, from some minimum value up to the value expected from the argument above. In other words, the energies of the proton and the electron don't add up to the initial rest energy of the neutron. To salvage the law of energy conservation, Fermi had to assume a third, unobserved particle carries away the missing energy.

The properties of the new particle could be easily deduced. The proton's positive electric charge and the electron's negative charge add up to zero, so the antineutrino must also have no electric charge—hence the name *neutrino*, or little neutral one. This is why the antineutrino was not observed—neutral particles don't leave tracks in a cloud chamber. The neutron has spin 1/2, as do the electron and proton. The quantum rules for spin declare that two spin 1/2 particles can't combine to produce spin 1/2, but the three spin 1/2 particles can do so. The neutrino must therefore have spin 1/2. (This also rules out the possibility that the missing energy is carried by photons, which have spin 1.) The mass of the neutrino can be inferred from the electron energies involved in beta decay. It was found to be nearly zero. The ever-present experimental uncertainties forbid the conclusion that the neutrino mass is exactly zero, but zero mass can be taken as a working hypothesis. Recent results (which will be revealed in Chapter 11) indicate that this hypothesis is probably wrong: Neutrinos most likely have mass, though it is much smaller than the electron mass. The particle in Fermi's theory is called an antineutrino rather than a neutrino to preserve yet another conservation law: the conservation of lepton number, which was previously discussed. Both electric charge and lepton number are conserved in beta decay. In fact, lepton number (and electric charge) conservation holds true in all reactions that we know of: strong, weak, and electromagnetic.

The strength of an interaction determines the time it takes for a particle to decay. The stronger the force, the faster the decay. Particles that decay via the strong force live about 10^{-24} seconds. In contrast, a particle that decays via the electromagnetic interaction lives about 10^{-16} seconds. This is 100 million times longer—a year compared to a blink of the eye. A particle decaying by way of the weak force typically lives about 10^{-8} seconds—another factor of 100 million longer. Weak though it is, life on earth would not be possible without this force. All life ultimately gets the energy it needs to exist from the sun. Plants trap the sun's energy by photosynthesis, and animals eat plants (or eat other animals that eat plants). Where does the sun get its energy? What keeps it shining? The answer: nuclear fusion reactions. (This means solar energy is actually a form of nuclear power—don't tell the environmentalists.)

One of the reactions that powers the sun involves the fusion of two protons. The fusion produces heavy hydrogen (a proton and a neutron stuck together), a positron, and a neutrino. The neutrino is a clue that the weak force is involved. A closer look at this reaction reveals what's really going on. One of the protons changes into a neutron, plus a positron and a neutrino—just the opposite of what happens in beta decay. The other proton is just a bystander (although it provides some of the energy needed to make the reaction go). Without this inverse beta decay process the sun would not shine and life would never have arisen on earth.

How to Design Your Own Universe

How is it possible for a neutron to turn into a bunch of other particles? Maybe the neutron is really a proton and an electron locked together, and its decay simply involves the particles escaping? This plausible explanation was quickly ruled out. The size of the neutron was known from scattering experiments (recall how the Nerf ball could be reconstructed from the scattered BBs). The energy required to bind a proton and an electron in such a small region would give the neutron a mass (via $E = mc^2$) much greater than what was actually measured. No, the neutron must really change its character in some fundamental way. By now, we are used to the idea of such transformations. Similar processes occur in QED and QCD.

In 1933, when the theory of the weak nuclear force was being born, no one had ever considered such a possibility. Fermi, in his audacity, claimed, at a time when Gell-Mann was four years old and QED was still a puzzling mess, that the neutron transformed into a proton, an electron, and an antineutrino, like an alchemist's mixture transforming into gold. Nowadays, physicists take this kind of transformation as a matter of course. In fact, all you need to do to invent a new theory of elementary particles is to decide which sorts of particles will transform into which other sorts of particles. The rest of the theory is determined by the structure of relativistic quantum field theory.

Here's how to design your own physics. Start by listing all the particles that you want to have in your universe, and give their masses, charges, and spin values. For instance, suppose we want to describe a universe that just has electrons, neutrinos, and photons:

Particle	Mass	Spin	Electric Charge
Electron	9.11×10^{-31} kg	1/2	−1
Neutrino	0?	1/2	0
Photon	0	1	0

The rules of relativistic quantum field theory will guarantee that we also have the corresponding antiparticles: positrons and antineutrinos. The photon is its own antiparticle. To turn this into a relativistic quantum field theory, we need to build the Lagrangian function that is used in Feynman's version of the least action principle. Two pieces are needed to build the Lagrangian: the propagator and the interactions. The propagator tells how a particle moves from point A to point B in the absence of interactions. It is completely determined by the particle's mass and spin. In Feynman diagrams, we simply draw a line. Keep in mind, though, that the line is shorthand for the sum of *all possible paths* the particle could take from A to B. All that's left to do to build our theory is to choose what interactions we want to have. We simply declare that whatever process we desire just happens—any particle can convert instantaneously into any other collection of particles. If we want electric charge to be conserved in our universe, we should choose interactions that respect the conservation law. In Feynman diagrams, the interactions are the junction points where several particle lines come together. In our example, let's let the electron interact with the photon.

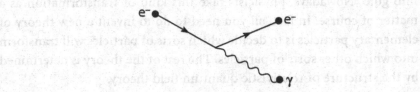

(This guarantees that all of QED will be included in our theory.) The neutrino has no charge, so we won't allow it to interact with the photon. Next, suppose we let the electron and the neutrino interact directly:

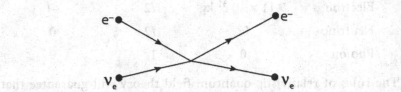

(In our simplified universe, this is the only role for the weak force.) Finally, we need to specify the *strength* of the interactions, or, equivalently, the probability for the specified transformation to occur. Each interaction diagram has an associated *coupling constant*, the number that specifies the strength of the interaction. (For particles with spin, there may be more than one coupling constant corresponding to the ways particles with spins in different directions interact.) In our example, there are two: the electron-photon coupling constant, and the weak coupling constant that specifies the strength of the electron-neutrino interaction.

That's all there is to it. Make up any particles you like, draw the diagrams for their interactions, and you have designed the physics (always excepting gravity) for your universe. You then apply the rules of relativistic quantum field theory to add up all the possible ways a process can take place. There's one snag to the whole thing, however. If our theory is to have any hope of describing the real world, it had better be mathematically consistent. That is, it needs to be renormalizable. When the calculations tell us the answer is infinite, there must be a way to sweep the infinities under the rug. For some choices of particle interactions, the theory is renormalizable; for others it isn't. With the electron-neutrino interaction that we chose earlier, there's no way to hide all the infinities that show up, and so no way to make the theory mathematically consistent. To get a theory that describes the real world, we'll have to do better.

The key to building renormalizable theories can be stated in a word: symmetry. Specifically, the Yang-Mills type of symmetry that we used to create the color theory of quarks and gluons is precisely what is needed for renormalizability. That's why the color theory is so useful (and will soon be revealed as an essential ingredient of the Standard Model)—it's renormalizable. Here's how it works: Quarks interact by exchanging a gluon. We say that the gluon *mediates* the strong interactions the same way the photon mediates the electromagnetic interaction. In contrast, the electron-neutrino interaction just proposed has no mediating particle. The electron and the neutrino just slam into each other.

In order to get a renormalizable theory, let's replace the previous electron-neutrino interaction with a Yang-Mills type interaction. Yang-Mills symmetry requires that the interaction be mediated by a new particle; let's just call it an X for now. In the basic interaction, an electron emits an X and turns into a neutrino, as in the top part of the illustration below. The bottom half of the diagram is just the same process in reverse.

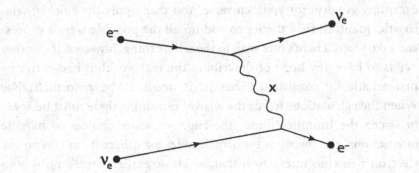

The X must have a negative charge, because the electron has a negative charge and the neutrino is neutral. The presence of the mediating X particle softens the electron-neutrino collision, in a sense. Instead of having the two particles just slam into each other, we now have the interaction proceeding by a gentlemanly exchange of the X. By adding this single interaction (and the analogous ones required by symmetry—for example, a positron emits the antiparticle of the X and turns into an antineutrino), we get a renormalizable theory of electron-neutrino interactions.

Let's try to build a realistic theory of the weak force using what we've learned so far. The particles that mediate weak interactions are called the W^+ and W^-. (Pronounced "dubya-plus" and "dubya-minus," these are each other's antiparticles and are collectively called the W particle.) We use the Yang-Mills trick for interactions between electrons and neutrinos. Yang-Mills symmetry requires that we introduce a third intermediate particle, known as the Z^0 (pronounced "zee-zero"). We can repeat the trick for the interactions of muons and neutrinos. Quarks, too, need to have weak interactions. The pion, for instance, decays into a muon and an antineutrino. We can include these easily enough: just decide what the coupling constants of the quarks with the W and Z^0 should be and include the appropriate Feynman diagrams. At this point, it might seem like we're done—we have a renormalizable theory of weak interactions. However, a complication crops up. The very Yang-Mills symmetry that guarantees renormalizability also forces the mediating particles to be massless. If they're massless, though, why had they never been observed? A massless W particle should be like a photon with charge—photons are easy to detect, and charges are easy to detect. No such particle had ever been seen in experiments. Moreover, a massless particle produces a long-range force (electromagnetism is the prime example), while the weak force acts only at subatomic distances.

(Gluons are massless intermediate particles, and you might be wondering why gluons don't suffer from the same objection. Recall that gluons are "colored" particles, and our study of the color force led to the conclusion that particles with color are never observed as free particles, what we called color confinement. For the same reason, there is no long-range color force: The gluons can't break free of the protons and neutrons in which they are confined.)

This was the problem in early 1971: The weak interaction needed a massless intermediate particle like the W for renormalizability, but the W needed to have mass to describe the weak interaction correctly. Most particle physicists were not thinking in these terms at that time, but a Dutch theorist, Martinus Veltman, was on the scent. A graduate student of his, Gerard 't Hooft, was just putting the finishing touches on the proof of renormalizability of Yang-Mills theories. Veltman called 't Hooft into his office, and this was the conversation[1]:

> M.V.: I do not care what and how, but what we must have is at least one renormalizable theory with massive charged vector bosons, and whether that looks like Nature is of no concern, those are details that will be fixed later by some model freak...

G. 't H.: I can do that.

M.V.: What do you say?

G. 't H.: I can do that.

't Hooft knew of a trick, called *spontaneous symmetry breaking*, which took a massless Yang-Mills theory, added an extra particle, and gave a new theory with intermediate particles that have mass (what Veltmann called "massive charged vector bosons"). What's more, 't Hooft had confidence based on his work on Yang-Mills theories that the resulting theory would be renormalizable.

Not only was he right about renormalizability, but it turned out that the "model freaks" had been there already. Veltmann learned of a 1967 paper by Steven Weinberg that described a Yang-Mills theory using the symmetry group SU(2) x U(1) (pronounced "ess-you-two-cross-you-one") that gave mass to the W particle via spontaneous symmetry breaking. It had been published and resoundingly ignored, as had an identical model developed independently by Abdus Salam, a theorist from Pakistan. According to Weinberg, from the paper's publication until 1970, it was cited in physics journals only once.

The reason for this neglect was that in 1967 no one knew that Yang-Mills theories were renormalizable, much less the theories complicated by spontaneous symmetry breaking. With 't Hooft's proof of renormalizability, the Dutchman had the first complete theory of the weak interactions. In 1972, the year after 't Hooft's paper, the Weinberg paper was cited 65 times, and the following year 165 times. As a bonus, the electromagnetic interactions were described by the same model. Weak and electromagnetic interactions had been unified into a single theory, which, as we will see, made some sharp, testable predictions, including the existence of three previously unsuspected particles. For their contribution, Weinberg, Salam, and Sheldon Glashow, who had worked on an earlier version of the unified theory, shared the Nobel Prize in 1979. Somewhat belatedly, Veltman and 't Hooft were awarded the prize 20 years later.

Breaking the Symmetry

What does it mean to break a symmetry? Imagine a pencil perfectly balanced on its tip on a table. There is a rotational symmetry to the situation: rotate the picture by any angle about the vertical axis of the pencil, and it remains the same. The slightest air motion will make the pencil fall, though. Now the original symmetry is broken: There is a difference between the direction parallel to the pencil and the one perpendicular to the pencil. Imagine an ant crawling along the tabletop. Before the pencil falls, the ant can roam at will (as long as it avoids the point where the pencil tip touches the table). After the fall, the ant suddenly notices a difference: it's easier to travel in one direction (parallel to the pencil) than in the other (perpendicular to the pencil, where the ant must climb over it). The symmetry breaking is spontaneous because any small push will cause the pencil to fall; the table doesn't need to be slanted for it to happen.

To see what this has to do with elementary particles, we have to leave Feynman mode and enter Schwinger mode: stop thinking about particles and start thinking about fields. According to Schwinger a quantum field can be thought of as an infinite collection of harmonic oscillators, one at each point in spacetime. Suppose now that there are two identical quantum fields at each point. The directions in which these fields oscillate are perpendicular to each other. (Think: two different tracks on which an ant could roller-skate.) This construction forms a bowl at each point in spacetime.

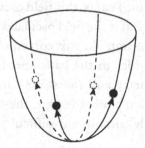

The bowl is symmetric, so these two types of oscillation are indistinguishable: The two particles (the quanta of the field oscillations) are identical.

Now suppose that the bowl has a dimple in the bottom.

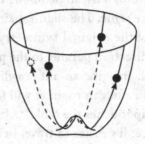

Since its shape resembles a sombrero, this is sometimes called the Mexican hat potential. When there is a lot of energy in the field, the dimple makes little difference, and the two directions are still identical. When there is little energy in the field, the situation is very different. As the energy is lowered, the field becomes trapped in the trough around the central hump (that is, it doesn't have the energy to get over the hump). Picture the average field value as a ball resting on the central hump. Like the pencil balanced on its point, the ball will roll off the hump at the slightest disturbance, and will end up at some arbitrary point of the trough. The exact location is irrelevant, since the trough itself is still symmetric. But now when the field oscillates about the ball's position, the same two directions look very different.

In the direction across the trough, oscillations occur just as if we had a harmonic oscillator. That is, this field oscillation looks like a normal particle with mass. But if the field oscillates the other direction, the trough is *flat*. A flat direction like this corresponds to a massless particle. Recall the skateboarder in the half-pipe: In the cross-pipe direction, there is no symmetry and the skateboarder oscillates back and forth; but in the direction along the pipe, the flat direction, the skateboarder moves smoothly without any difficulty.

To sum up: We started with two identical fields and a potential that is symmetric. Oscillations of the field at high enough energy to be above the hump in the potential are identical—the two particles represented by the field are indistinguishable. At low energy, though, the field must oscillate in the trough, where the original symmetry is broken. One of these oscillations represents a massive particle, the other a massless particle. Spontaneous symmetry breaking has transformed two identical particles into two apparently very different particles.

Now back to the Weinberg-Salam theory of electroweak interactions. Start with an SU(2) x U(1) Yang-Mills theory. It has four massless intermediate particles, similar to the gluons in QCD. Now add two new identical particles, together with the Mexican hat potential. These new particles are called Higgs particles, after the Scottish physicist Peter Higgs, who first introduced spontaneous symmetry breaking into elementary particle theory. Finally, let the Higgs particles interact with all of the other particles in the theory.

At low energy, spontaneous symmetry breaking occurs. Using the Schwinger picture, the quantum field associated with the Higgs particle looks like a Mexican hat at each point in space. When the energy in the Higgs field gets lower than the hump of the Mexican hat potential, the Higgs field rolls off the hump and ends up in the trough. This subtle shift occurs *at every point in the entire universe*. Now all the interactions between the particles cause something very odd to occur. We expect to end up with one massless Higgs and one massive Higgs particle as described earlier. Instead, we find only one massive Higgs particle—the other Higgs particle disappears from the theory completely. What's more, three of the four intermediate particles become massive, while the fourth remains massless. This is just what we wanted. The massless particle is the photon, and has exactly the right properties for the electromagnetic interaction. Because of their interactions with the Higgs, the other three particles get masses. Physicists like to say that the intermediate particles "eat the extra Higgs particle" and gain weight. What's actually happening is this: The Higgs Field got shifted when it rolled down into the trough. The intermediate particles are connected to the Higgs by the interactions we gave them. Wherever they go, they encounter the shifted Higgs field. Like someone trying to run through waist-deep water, the intermediate particles have to move against the drag of the Higgs field. The now-massive intermediate particles are the W^+ and W^-, and the Z^0. This was the accomplishment that won two Nobel prizes: on the one hand, the intermediate particles have picked up the mass that is necessary to explain the short-range nature of the weak force; on the other hand, because the masses arise from spontaneous symmetry breaking, the theory remains renormalizable.

The masses of the W and Z^0 particles could be calculated from the theory, but they were beyond the reach of the particle accelerators of the early 1970s, so it was not surprising that the particles had never been detected. Could they be found? The theorists had thrown the gauntlet, and experimentalists were not slow to take up the challenge. The Italian Carlo Rubbia convinced CERN, the European particle physics laboratory, that the existing proton accelerator could be converted to a machine that stored both protons and antiprotons. The antiprotons would travel the opposite direction of the protons, and when the beams were brought together, the resulting head-on collisions between protons and antiprotons would have enough energy to produce the massive intermediate particles. The Europeans were in a race with Fermilab in Illinois, which was struggling to get its Tevatron accelerator functional. A win would mean a lot to the Europeans, who had lagged behind the United States in high-energy physics for almost 50 years. Rubbia delivered, and by January 1983, the discovery of the W was announced. The Z^0 was found later the same year. Both masses were just where they were supposed to be. Rubbia and Simon Van der Meer shared the Nobel Prize in 1984.

Finally, and surprisingly, the Higgs particle gives mass to the electrons and (if you include them in the model) the quarks. Since the Higgs interacts with these particles, they experience the same sort of drag effect that gave mass to the W and Z^0. That is, starting with a theory of entirely massless particles—massless electrons, massless quarks, massless neutrinos, and massless intermediate particles—and adding in the Higgs particle with its Mexican hat potential, spontaneous symmetry breaking yields a theory with massive electrons, massive quarks, massive W^+, W^-, and Z^0 particles, and a massless photon. Just what we see in the real world. Leon Lederman calls it the God Particle: Without the Higgs there would be nothing in the universe but massless particles. A massless electron could never be captured by a proton to form an atom. Without atoms, no galaxies, no stars, no earth would form. The universe would be very boring. Only by breaking the symmetry do we get massive particles that can become the constituents of matter.

The amazing electroweak unification was accomplished at this cost: Weinberg and Salam needed to invent four new particles. The three massive intermediate particles have since been found, with the masses and properties predicted. The remaining particle predicted by unification is the Higgs particle, which has not yet been detected in any experiment. Is its mass simply too large for the reach of today's accelerators? Or is there some deeper problem with the Standard Model?

In August of 2000, physicists working at the Large Electron Positron Collider (LEP) at CERN saw a few events that looked like Higgs events are expected to look. LEP was due to be shut down in order to be upgraded to much higher energy. The CERN directors faced an agonizing decision: to delay the upgrade in order to look for enough Higgs-like events to declare that they had discovered the elusive particle, or to shut down, perform the upgrade, and look for the Higgs with the improved machine. After a one-month delay during which additional events failed to materialize, the decision was made to do the upgrade. Scheduled to begin operation in 2007, the new Large Hadron Collider will reach beam energies of several trillion electron-volts. The mysteries of the Higgs may soon be solved.

Chapter 10

The Standard Model, At Last

Everything important is, at bottom, utterly simple.
—John Archibald Wheeler

" I want to know how God created this world. I am not interested in
this or that phenomenon, in the spectrum of this or that element.
I want to know His thoughts; the rest are details," said Albert Einstein.[1]
To know God's thoughts, the laws by which the universe runs—that is
the way physicists view their goal. The details, the particular masses of
individual particles, are important, but only as a means to a wider goal.
Physicists want to know about the general principles, the overarching
structure, the architecture of the universe.

During the course of the twentieth century, physicists painstakingly
deduced bits of that structure. At the core is the relativistic quantum
field: the idea that all interactions are fundamentally an exchange of particles, that these exchanges are completely random and unpredictable,
and that we can discover only the likelihood of any process.

The Standard Model is built from relativistic quantum field theories we already know about. QED is incorporated into electroweak
theory, and the whole is woven together with QCD to create a single
theory whose essential elements can be written in a single equation.
This equation is the simplicity at the bottom of it all, the ultimate
source of all the complex behavior that we see in the physical world:
atoms, molecules, solids, liquids, gases, rocks, plants, and animals.

In this chapter, we finally reach our goal: the Standard Model of elementary particles. It is not, of course, the ultimate goal. Gravitational interactions are left out, and without them we certainly can't lay claim to a full understanding of the universe. Still, the Standard Model is an incredible achievement—a single theory that not only summarizes everything we know about matter and its interactions, but also answers fundamental questions about the symmetry and structure of the universe.

In the days of the Eightfold Way, only three flavors of quark—up, down, and strange—were needed to explain all of the then-known particles. By 2000, it was known that there are actually six quark flavors, and that they come in pairs, or families. One quark in each family has charge +2/3 and the other has charge −1/3. The masses also follow a clear pattern: The quarks in the second family are more massive than those of the first family, and those of the third family are more massive still.

The leptons, the lightweight particles, come in pairs as well. The electron interacts with its neutrino. The muon, as we have seen, is an overweight version of the electron. It has the same charge, the same spin, and has weak interactions but no strong interactions, just like the electron. The muon interacts with a second type of neutrino, known as the mu neutrino. Muons ignore electron neutrinos entirely, and electrons likewise ignore mu neutrinos. A third lepton, the tau, completes the pattern. The tau has a larger mass than the muon, but has the same charge and spin, lacks any strong interactions, and has its own associated neutrino. The existence of the tau neutrino was only experimentally confirmed in July 2000, although physicists had long been confident it would be found. It is the last but one of the predicted particles of the Standard Model to be found—only the Higgs remains.

Physicists thus found that Nature repeated herself, producing both quarks and leptons in three families.

	Family		
	1	**2**	**3**
Leptons			
Neutrinos	v_e (electron neutrino)	v_μ (mu neutrino)	v_τ (tau neutrino)
Electron and relatives	e^- (electron)	μ^- (muon)	τ^- (tau)
Quarks	u (up)	c (charm)	t (top)
	d (down)	s (strange)	b (bottom)

This chart gives us a new sort of periodic table: the periodic table of the fermions. Only the particles of the first family appear in the atoms that make up the matter around us and in us. The other particles appear only fleetingly, the short-lived by-products of high-energy collisions.

It is striking that the number of quark and lepton families is the same. This is actually a crucial feature of the Standard Model. If there were different numbers of quark and lepton families, the model wouldn't be renormalizable. The Standard Model is silent about why the fermions enjoy this family relationship, or why the number of families should be precisely three. A happy coincidence, or is there some deeper reason?

Now all the ingredients for a Theory of Almost Everything are in place. Quarks and gluons interact by way of QCD. Electrons, photons, and neutrinos interact by way of the unified electroweak theory. All that's left is to cobble the two theories together into one relativistic quantum field theory that describes all the interactions of all the particles ever observed, plus the still unobserved Higgs.

According to the recipe for designing your own physics, begin by listing the particles of the theory. In addition to the three fermion families just listed, we need the following particles:

<u>Intermediate Particles</u>   W Z^0

                              γ (photon)   g (gluon—comes in
                                                             8 varieties)

<u>Symmetry breakers</u>   H (Higgs doublet)

Of course, each particle has its antiparticle (except for those that are their own antiparticle). I won't bother to list those.

Next, for each particle, we need the propagator that describes how the particle gets from place to place in the absence of interactions. The propagator is entirely determined by the mass and spin of the particle. In Feynman diagrams, we draw a straight line for the spin $1/2$ fermions, a squiggly line for the spin 1 intermediate particles, and a dashed line for the spin 0 Higgs.

Finally, we need the interactions. If we had to consider all possible interactions of the 37 particles in the table, we'd be in trouble. Fortunately, the couplings of the intermediate particles with each other are fixed by the symmetries of Yang-Mills theory. The quarks, of course, interact with the gluons as required by color symmetry, and all the fermions (electrons, neutrinos, and quarks) interact with the intermediate particles of the electroweak theory (the W and Z^0 particles, and the photons). The Higgs, as we saw in the previous chapter, interacts with almost everything. It interacts with the intermediate particles, the leptons, the quarks, and, crucially, with itself, in just the right manner to produce the Mexican hat potential. These are the interactions that give all of the particles their mass after symmetry breaking occurs. If we knew the coupling constants for these interactions, we would know the masses. Unfortunately, the Standard Model tells us nothing about these coupling constants. The quark and lepton masses can't be predicted from the Standard Model; they must be determined experimentally. (For more details of all these interactions, turn to Appendix C.)

There you have it. Knowing the propagators and the interactions of the Standard Model you can, in principle, calculate the probability of any process in the universe. (Well, any process not involving gravity.) Why quarks stick together in protons, why atoms bind into molecules, why my feet don't pass through the floor beneath them—it's all there. In principle. In practice, the calculation requires knowing how to translate these diagrams into equations. Below, in simplified notation, is the starting point: the Lagrangian function that summarizes all of the propagators and interactions we just listed.

$$L = -\frac{1}{4}G^2 - \frac{1}{4}W^2 - \frac{1}{4}F^2 + (\nabla H)^2 + \mu^2 H^2 - \lambda H^4 + i\sum_j \bar{f}_j \nabla f_j - \sum_{jk} c_{jk}\, \bar{f}_j H f_k$$

This one equation describes everything we know about the way matter behaves, except, of course, for gravity. In the equation, G describes the gluon field and its interactions, W stands for the SU(2) field (describing the W^+, W^-, and Z^0 particles and their interactions with each other), F is the U(1) field, H is the Higgs, and f_j is shorthand for all the fermions: electrons, neutrinos, and quarks.

There are 18 adjustable numerical parameters in total in the Standard Model. Think of the Standard Model as a machine with 18 knobs. If we twiddle the knobs just right, the machine spits out nearly perfect predictions about any (nongravitational) process in the universe. Just what are these magical numbers that encode everything we know about physics? First, the strengths of the color force and the electroweak force. Second, the Higgs couplings to the quarks and leptons that give them mass. Third, the numbers that specify the shape of the Mexican hat potential. Finally, there are a few additional parameters having to do with the interactions of the quarks: these are discussed in Appendix C.

Despite the beauty of the Higgs mechanism for spontaneous symmetry breaking and its spectacular success in predicting the W and Z^0 masses, it must be admitted that our Standard Model machine, with its 18 knobs, is a bit unwieldy. Weinberg himself called it a "repulsive" model in 1971, and theorist Thomas Kibble remembers reading Weinberg's paper and coming to the conclusion that, "It was such an extraordinarily ad hoc and ugly theory that it was clearly nonsense."[2] Is there rhyme or reason to the 18 parameters, the three generations of fermions, the odd combination SU(3) x SU(2) x U(1)? We'll return to this question in Chapter 12.

The Standard Model summarizes and organizes everything we know about the particles that are the building blocks of our world. It makes testable, highly accurate predictions about reactions such as the nuclear reactions that cause the sun to shine. Apart from predicting the W and Z^0 particles, which after all can only be produced using specialized machines costing hundreds of millions of dollars, what does the Standard Model tell us about the universe?

The Universe Is Left-Handed

We could call it the Lewis Carroll question: Is the looking-glass world different from our world? When Alice went through the mirror, she was surprised to find a very different world on the other side. Physicists were likewise surprised to find in 1956 that the mirror world is not, in fact, identical to ours. Everyday physical processes look essentially the same in the mirror world as in our world. If you filmed a game of billiards by pointing the camera at a mirror instead of directly at the table, the film would not reveal the trick. Right-handed players would appear to be playing left-handed, and vice versa. Nothing about the movement and collisions of the balls would look at all unusual. The same can be said of electromagnetic interactions or gravity. Until 1956, physicists had never encountered a process that was not also possible in the mirror world.

The process that first showed a violation of this mirror symmetry, or parity as it is called, was our old friend beta decay, in which a neutron decays into a proton, an electron, and an antineutrino. Ignoring the hard-to-detect antineutrino, here's how the experiment might look in our world and in the mirror world.

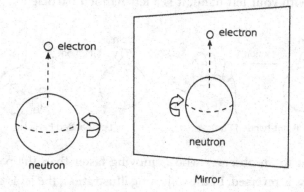

The mirror world experiment looks just the same, except that the spin of the neutron is reversed in the mirror world. If the electron is coming from the "north pole" of the neutron in our world, then it is coming from the "south pole" in the mirror world. Now, if parity was a symmetry of beta decay, these processes would have to be equally likely. A careful experiment done by Chien-Shiung Wu of Columbia University (using cobalt-60 rather than free neutrons) proved that north pole electrons were slightly more likely than south pole electrons. Mirror symmetry was violated in weak interactions. Wu's experiment was quickly followed by others that confirmed parity violation, and in 1957, the two theorists who had suggested the possibility, Tsung-Dao Lee and Chen Ning Yang, became the youngest Nobel Prize winners ever, at the ages of 30 and 34.

By the time the Standard Model was being put together in the early 1970s (based, as we have seen, on another contribution of Yang's, the Yang-Mills symmetry), parity violation was a long-accepted fact, and was therefore built into the theory in a very blatant way: All the neutrinos are left handed. The *handedness* of a particle is defined by its spin and its direction of motion. If you curl the fingers of your right hand in the direction of the spin and your thumb points in the direction of the particle's motion, it is a right-handed particle. If you can do the same with your left hand, it is a left-handed particle:

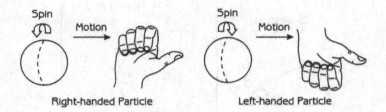

Right-handed Particle   Left-handed Particle

Notice that to an observer who is moving faster than the particle, the handedness is reversed. In the following illustration, the jet is overtaking a left-handed particle. From the jet's point of view, the particle is moving to the left. So, in this reference frame, the particle is right-handed.

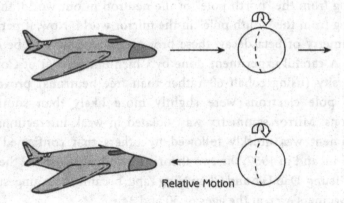

How is it possible that the Standard Model has only left-handed neutrinos? Remember that neutrinos have no mass in the Standard Model, so they are always traveling at the speed of light. The jet can't go faster than the speed of light, so it can never overtake the neutrino. There is no reference frame in which the neutrino looks right-handed. Later, we will consider how the Standard Model must be modified if neutrinos actually have mass.

Just three days before Wu announced her parity-violation results, Wolfgang Pauli wrote to a friend, "I do not believe that the Lord is a weak left-hander."[3] For once his intuition was wrong. The looking-glass world is not the same as our world.

Physics of the Large and the Small

The Standard Model contains three families of quarks and three families of leptons. Are there other, heavier families still to be discovered? Surprisingly, we can answer this question by studying the decays of the Z^0. The Standard Model predicts that the Z^0 can decay into a neutrino-antineutrino pair of any flavor. The more families there are, the more possible routes for the Z^0 to decay. Every extra decay route increases the probability that the Z^0 will decay, and so decreases its expected life-time. So, a measurement of the lifetime of the Z^0 tells us about how many fermion families there are. The answer is that there can be no more nor less than three families. It remains a mystery why nature cloned the electron family at all. But, if there are more fundamental particles yet to be discovered, they aren't simply further duplicates of the families we already know.

Particle physicists weren't surprised to learn from Z^0 decays that there were only three lepton families; they had already been told the answer by the astrophysicists. To understand how the study of the skies got to the answer before the big accelerators did, we need to know a few things about the big bang origin of the universe. The motivation for the big bang model comes from the fact that everywhere we look galaxies are moving away from our galaxy. The farther the galaxy, the faster it is racing away. Now imagine running the movie of the universe backwards. At an earlier time, all the galaxies were closer together. Earlier yet, and there were no galaxies; a uniform hot gas filled the universe. About 93 percent of the atoms in the gas were hydrogen and about 7 percent were helium, with trace amounts of other elements: Those are the percentages we measure in the leftover gas clouds between galaxies today. Running the movie even farther back, the gas filling the universe becomes so dense and hot that atoms cannot exist; they are broken up into individual protons, neutrons, and electrons. We have reached a time only 1/1000 of a second after the big bang itself.

Still earlier, and the temperature is so high that neutrons and protons are torn apart into individual quarks. At this time, the entire currently observable universe, all the stars and galaxies that we can see, is packed into a ball that could fit comfortably between the current positions of our sun and the nearest stars. We have reached the quark era, when the universe was less than 10 microseconds old. The entire universe is filled with an incredibly hot, dense quark soup. All of the processes of the Standard Model are occurring at a frantic pace: Energy is constantly being converted from quarks to photons to electrons to neutrinos to Ws and Zs. Since particles of all types will be produced, the presence of extra particle families affects the energy balance of all the other particles. (Actually, it is only the presence of extra light particles, like neutrinos, that matters, since these are the easiest to produce.) As a result, changing the number of fermion families changes the numbers of neutrons and protons that are produced when the universe cools past the quark era. The neutron/proton ratio in turn affects the percentages of hydrogen and helium that get made when the universe cools still further. In this way, a fundamental fact of observational astronomy, the abundance of hydrogen and helium in the universe, is connected in a crucial way with a fundamental fact about the Standard Model, the number of fermion families. The astrophysical evidence implied the existence of only three fermion families even before measurements of the Z^0 lifetime led to the same conclusion.

At this point, I need to make a confession. There is a way to avoid the conclusion that the three families that we know are the only ones that exist. The Standard Model assumes that all neutrinos are massless, like photons. Recent experiments show, though, that this assumption is almost certainly wrong. These experiments take us beyond the Standard Model, however, so I leave discussion of them to Chapter 11. But both of the arguments "proving" that only three families exist rely on the assumption that neutrinos are, if not strictly massless, at least very much lighter than an electron. So we can't rule out the possibility of a fourth family with a heavy neutrino.

The convergence of results from astrophysics and elementary particle physics is part of a surprising synergy that has developed in the past 20 years where the physics of the very large (the universe, galaxies, supernovas) and the very small (the Standard Model) come together to produce new results. The Standard Model brings us full circle, to a point where particle physics and astrophysics are no longer separate disciplines. New ideas in one area can have a profound impact on the other area. This was certainly the case when a young particle physicist at SLAC developed a twist on the big bang model known as the inflationary universe.

Blowing Up the Universe

In 1979, Alan Guth was a postdoctoral fellow still looking for a tenure-track job. Together with Henry Tye, he began to consider what effect the Higgs field would have on the picture of the early universe derived from the big bang model. The effect, he discovered, was profound, and solved at one blow many of the outstanding difficulties of the big bang model. (I will describe an improvement of Guth's model, the new inflationary universe model, developed by Andrei Linde and independently by Paul Steinhardt, rather than Guth's original theory.)

In the very early universe, the temperature must have been so high that spontaneous symmetry breaking could not occur. The Higgs field had enough energy to keep it above the hump in the Mexican hat potential. If we represent the average value of the Higgs field by a ball rolling in the "bowl" (the potential), it starts off perched on top of the hump. As the universe expands and cools, the energy in the Higgs field drops below the top of the hump. Now the situation is unstable, like the pencil balanced on its point. The ball wants to roll down into the trough. Since the cooling happens so quickly, though, it remains momentarily at the center, creating a situation known as supercooling. A similar situation occurs when very pure water is cooled below its freezing point. Not having any impurities to start the formation of ice crystals, the water remains liquid until it is disturbed by some external influence, at which point it all freezes very rapidly. In the supercooled early universe, the Higgs field is disturbed by quantum fluctuations and the ball begins to roll off the hump. Here's where things start to get interesting. According to Einstein's theory of general relativity, the rate of expansion of the universe depends on the state of the matter and energy in the universe. Guth discovered that the transition out of the supercooled state caused an exponential expansion of the universe. The size of any given region of the universe would double in the unimaginably tiny time of 10^{-37} second. The universe might go through 100 such doublings in the first moments of its existence. At the end of this inflationary period, the currently observable universe would have been only about a 100 miles across.

The word "about" in the previous sentence should be taken very liberally. The predicted size depends on many assumptions and details about the inflationary model chosen. The exponential expansion is so great, though, that we can say with confidence that *before* inflation, the entire currently observable universe—all the stars, gas, and dust, not only in our galaxy, but in all the galaxies from us to the farthest visible quasars—was crammed into a space much smaller than a proton.

Finally, after the ball rolls off the hump, the field oscillates in the trough of the Mexican hat, and the oscillations gradually diminish as the remaining energy in the Higgs field gets converted to all of the various Standard Model particles. Eventually, the energy level drops until the field value comes to rest in its new stable position. The particles created from the Higgs energy go on to become the protons, neutrons, electrons, and photons that make up everything in the universe. This gives a new meaning to Leon Lederman's term for the Higgs: the God Particle.

Guth, followed by Linde and Steinhardt, went on to show how inflation solved several puzzles of the standard big bang scenario. Why is the observable universe so uniform? Because it started out smaller than a proton, and such a small region must have been at a uniform temperature. Why is the universe so nearly flat? Any curvature would have been smoothed out by the expansion. Why have magnetic monopoles never been observed? (Magnetic monopoles—magnetic particles with a "north" but no "south" pole or vice versa—are predicted by grand unified theories, which we will encounter in the next chapter.) Any monopoles that existed before inflation would be so spread out by the expansion that we would have very little chance of encountering one.

Attractive as it is for solving cosmological difficulties, inflation has its own problems. More modern views of the inflationary universe require the hump in the Mexican hat potential to be low and flat. Exactly how flat it needs to be depends on the details of the theory under consideration, but it often ends up so flat that it is useless for spontaneous symmetry breaking. Some physicists have suggested that inflation is not caused by the Higgs boson at all, but by some other particle with a Mexican hat potential, which they call the inflaton. This of course raises more questions: Why is there an inflaton? Why is its potential so flat? Answers will come, if at all, only from theories that go beyond the Standard Model.

Is it unconscionable hubris for physicists to claim they know what the universe was like 14 billion years ago, a fraction of a second after its creation? The answer hinges on another question: how well do we know the limits of our theory? The early universe discussion relied on the coupling of the big bang model of general relativity with the Standard Model. The Standard Model works flawlessly (except for those neutrino masses) at the energies attainable at current accelerators—up to about 200 billion electron-volts. In the big bang model, this is the average energy about a billionth of a second after the big bang. We have no reason to doubt the predictions of the Standard Model at these early times.

How about general relativity? Its fundamental parameters are the gravitational constant, G, and the speed of light, c. No one knows the limits of validity of general relativity; no experiment has ever been done that is in conflict with the theory, but we might expect that it will break down when quantum effects become important. We can combine the constants G and c with Planck's constant, $\hbar$, from quantum

$$E_{pl} = \sqrt{\frac{\hbar c^5}{G}} \approx 10^{19} \text{ billion electron volts}$$

mechanics to get an energy:

If this energy, called the Planck energy, is truly the limit of validity of general relativity, then we should be able to trust the big bang model as early as the Planck time: 10^{-43} seconds after the big bang.

As far as we know, then, we are working within the limits of validity of the theories when we describe the early universe. What's more, we can make predictions (although they are really *retrodictions* about things that happened long ago) using this picture: the primordial abundance of deuterium, helium (^3He and ^4He), and lithium, all of which accord well with observations. The Standard Model–big bang model combination allows us to travel in our imagination confidently back to the first milliseconds of the universe's existence.

Any physical theory has its limits. Maxwell's equations work beautifully to describe electromagnetic phenomena, but only up to a certain point. When dealing with the physics of the very small, electrons in atoms or photons in high-energy scattering experiments, Maxwell's equations fail; they must be replaced by the equations of the Standard Model. Sometimes, it is a new experiment that hurls the experimenter beyond the limits of validity of the theory, as when the cosmic ray experiments of the 1950s and 1960s began turning up unexpected new particles. Sometimes, though, it is the theory itself that tells us its limitations. If we use the equations of general relativity to push the big bang model back past the Planck time, we find that the temperature everywhere in the universe rises toward infinity. Surely, this is a sign that the theory has broken down, that it is incapable of making correct predictions in that extreme environment so far beyond our experience.

Does the Standard Model give us any hints about its limits of validity? A careful look at the behavior of the three forces in the theory reveals a surprising fact. Remember that the strength of a force depends on the energy scale of the particular experiment. The strength of the electromagnetic force increases as you go to higher energies, but the strength of the strong force decreases. The weak force likewise changes with the energy scale. Plotting the strength of the force as a function of the energy, we find that the three curves meet very nearly at a single point (α_1 is the electromagnetic coupling constant, α_2 is the weak coupling constant, and α_3 is the strong (color) coupling

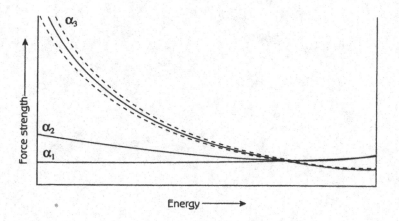

constant).

It appears that the curves meet at energy scale of around 10^{15} GeV. Is this pure coincidence? Or is Nature trying to tell us that the three forces are really different aspects of a single, unified force that our accelerators are just too feeble to see? This unification energy scale is a hundred trillion times larger than current accelerators can reach; as Leon

Lederman says, this "puts it out of the range of even the most megalomaniacal accelerator builder."[4] However, theories that unify the three forces at this energy also leave traces at lower energies, traces that might show up soon in accelerator experiments or through astronomical observations. Let's take a look at some experiments that might already be showing us where the Standard Model breaks down.

Chapter 11

The Edge of Physics

Birdboot: Where's Higgs?
—Tom Stoppard, *The Real Inspector Hound*

The Standard Model is by far the most successful scientific theory ever. Not only have some of its predictions been confirmed to spectacular precision, one part in 10 billion for the electron magnetic moment, but the range of application of the theory is unparalleled. From the behavior of quarks inside the proton to the behavior of galactic magnetic fields, the Standard Model works across the entire range of human experience. Accomplishing this with merely 18 adjustable parameters is an unprecedented accomplishment, making the Standard Model truly the capstone of twentieth-century science.

Why are physicists not content to rest on their laurels and accept general relativity (for gravity) and the Standard Model (for everything else) as the ultimate theories explaining all the interactions in the universe? The answer is two-fold: On one hand, the theories themselves hint at new physics beyond the Standard Model, and on the other, recent experiments are starting to require it.

Where Have All the Neutrinos Gone?

The first indication of physics beyond the Standard Model came from a surprising direction. It was not the multibillion dollar accelerators that produced the evidence, but an experiment that set out to check an apparently well-understood phenomenon—the physics of the sun.

A complete understanding of the sun involves the difficult interaction of hot, flowing plasma with the sun's magnetic fields, as well as the delicate balancing of the outward pressure of the hot plasma with the inward pull of gravity. The nuclear reactions that produce energy in the sun's core, on the other hand, were thought to be well understood. Many of those reactions produce neutrinos. Since neutrinos interact so weakly with other matter, they escape the sun's core without any alteration, providing a glimpse of the conditions in the sun's interior. An experiment designed by Raymond Davis and his collaborators detected a tiny fraction of the solar neutrinos that constantly rain onto the Earth. In 1968 they reported their results: only about half of the expected number of neutrinos.

Particle physicists viewed this result with interested skepticism. The question: Was there a problem with the physics of the nuclear reactions (the Standard Model did not exist yet), or with the physics of the sun, the so-called standard solar model? As the Standard Model was developed and verified through the 1970s and 1980s, particle physicists' confidence grew, and many assumed that the solar physics would be found in error. In experiment after experiment, though, the solar neutrino deficit persisted. For 35 years, increasingly sophisticated experiments looked for neutrinos from a number of different solar reactions and, slowly, a consistent picture emerged. The *solar* physics was fine, but the neutrinos were behaving strangely. According to the Standard Model, all of the neutrinos produced in the sun should be electron neutrinos. However, the experiments clearly showed that some of these neutrinos were changing their flavor during their trip to earth—they were oscillating from electron neutrinos into some other type. Presumably, that other type of neutrino is a mu or tau neutrino, but it is possible that the electron neutrinos are transforming into some as yet unknown neutrino type.

The theoretical explanation of these neutrino oscillations requires that at least one of the three neutrinos have a mass. Since all neutrinos are massless in the Standard Model, these experiments, the implications of which are just becoming clear in the first years of the twenty-first century, are truly the first indications of physics beyond the Standard Model.

Can the Standard Model be modified to include neutrino masses? As we've seen, a particle with mass cannot be purely left-handed, as neutrinos are in the Standard Model. At the very least, then, we need to add right-handed neutrinos to the theory. With these in hand, we can proceed to make neutrinos massive by the same trick that made the electron and quarks massive, by coupling to the Higgs field. Spontaneous symmetry breaking then leads to neutrino masses. All that's needed is to adjust the couplings to make the masses come out right.

This simple-minded approach has some difficulties. Experiments have shown that neutrino masses must be at least a million times smaller than the electron mass. Why should the coupling between the Higgs and the neutrino be a million times less than the coupling between the Higgs and the electron? Our model gives no way of answering this question. More seriously, the right-handed neutrinos we introduced count as additional light neutrinos, which are ruled out by accelerator experiments and by astrophysical considerations. As we will see in the next chapter, a class of theories known as grand unified theories (GUTs) allows massive neutrinos to be included in a more natural way.

An experiment now running at Fermilab, called the Mini Booster Neutrino Experiment (Mini-BooNE), is expected to produce results sometime in 2005. Mini-BooNE examines a beam of mu neutrinos to see if they transform into electron neutrinos. If this oscillation is seen, a second phase of the experiment (BooNE) will aim for a precise measurement of the transformation rate. A second Fermilab experiment, called the Main Injector Neutrino Oscillation Search (MINOS), uses a pair of detectors, one at Fermilab itself and one 430 miles away in a mine in Minnesota, to look for transformations of mu neutrinos into tau neutrinos. Taken together, Mini-BooNE and MINOS should give a very clear picture of neutrino oscillations. One goal of these investigations is to clarify the question of neutrino masses. The transformation rate of one neutrino flavor into another is directly related to the difference in the two masses. As a result, the oscillation experiments won't be able to determine the actual masses, only the pattern of mass differences. Of course, any neutrino masses other than zero take us beyond the Standard Model.

An even more intriguing possibility is suggested by the results of an experiment called the Liquid Scintillation Neutrino Detector (LSND) that finished a six-year run in 1998. Like the experiments that looked at neutrinos coming from the sun, LSND also provided evidence of neutrino oscillations. However, the rates of oscillation measured in the various experiments are very hard to reconcile unless one postulates the existence of a *fourth* type of neutrino. Now, we saw in the previous chapter that the decay rate of the Z^0 constrains us to exactly three light neutrinos. This conclusion can be avoided if the fourth neutrino doesn't interact at all with the Z^0. The symmetries of the Standard Model (extended to include the extra neutrino) then forbid the new neutrino from interacting with any of the other particles except the Higgs. For this reason, these hypothetical neutrinos are called sterile. If this fourth neutrino exists, it must be of a very different ilk than the three neutrinos we already know of. It can't be a member of a fourth family like the other families in the Standard Model. It would be a complete loner, a particle with no relatives and almost no interactions. If it could be proven that a sterile neutrino exists, a whole new chapter in elementary particle physics would begin.

The Scum of the Universe

I have been calling the Standard Model the Theory of Everything Except Gravity. However, a collection of surprising observations, which

I have been ignoring until now, indicates that the Standard Model is not even close to being the whole story. About 85 percent of the matter in the universe cannot be accounted for by the particles of the Standard Model. It is astronomy once again, rather than accelerator physics, that forces us to this astonishing conclusion. This result comes not just from a single observation but from a variety of different astronomical observations, providing the sort of converging lines of evidence that are crucial to acceptance of an idea as a scientific fact.

The stars in a galaxy orbit the galactic center just as the planets orbit the sun in our solar system. Newton's inverse-square law of gravity (or general relativity—the theories are nearly identical for galactic scales) predicts that the speed of a star will fall off as you move away from the galactic center. Actual observations tell a different story. For many galaxies, the speed of stars we observe is nearly constant, instead of falling off as expected. The discrepancy disappears if we assume there is more mass in the galaxy than we can see. Astronomers call the mysterious extra mass *dark matter*; we don't see it, but we can detect its presence by its gravitational effects. The amount of dark matter, deduced from the observed star speeds, far exceeds the amount of visible matter—stars, planets, gas, and dust—in the galaxy. Other observations support this astonishing conclusion. Measurements of the relative speed of pairs of galaxies, of speeds in galactic clusters, and of gravitational lensing (the bending of light as it passes by a galaxy) combine to fix the amount of excess mass at about six or seven times the amount of normal matter.

What makes this discovery so shocking is the fact that most of the dark matter must be something other than the protons, neutrons, and electrons that make up ordinary matter. That is, it cannot be dead, dark stars, planets too small to detect, or loose gas and dust that make up the dark matter. Dark matter has never been directly detected: How, then, can we know that it's not normal matter? The answer comes from the detailed picture of the early universe that the combined big bang–Standard Model provides. The production of the light elements, deuterium, helium, and lithium, in the first few minutes after the big bang is very sensitive to how many protons and neutrons are around. If the dark matter were normal matter, there would have been seven times as many protons and neutrons around during the first moments of the big bang. But then the light elements would have been produced in much greater quantities than we observe. Since protons, neutrons, and electrons are the only stable massive particles in the Standard Model, dark matter takes us beyond. We are forced to conclude that the bulk of the matter in the universe is something mysterious, something not included in the Standard Model. All of the galaxies we see, with all of their stars, planets, and dust clouds, are only a sort of scum on the fringes of enormous, invisible clouds of dark matter. Or, as physicist Pierre Ramond puts it more poetically, "we should think of luminous matter as the foam that rides the crest of waves of the dark matter in the cosmic ocean."[1]

What might this mysterious matter be? Observations indicate that our own galaxy is full of dark matter, too. We are presumably surrounded by it at all times, yet we have never detected it except by observations of distant galaxies. It must therefore be something that interacts only very minimally with ordinary matter. It doesn't radiate light or scatter it, so it must have no electric charge.

One possibility immediately springs to mind: massive neutrinos. They are, of course, neutral, and they interact only weakly with other matter. The universe is presumably full of neutrinos left over from the big bang. That, at least, is what the combination of the Standard Model and the big bang model predicts. Neutrinos have a big disadvantage as dark matter candidates, though. Their mass, if not zero, is very small. This means that in the early universe they were whizzing around at speeds near the speed of light (for which reason they are considered "hot" dark matter), a condition that seems to interfere with galaxy formation, according to computer simulations. We need to look elsewhere for dark matter.

Hypothetical dark matter particles that are much more massive than the proton are known as weakly interacting massive particles, or WIMPs, for short. In some supersymmetric theories, which we will learn about in the next chapter, there is a stable, neutral particle called the *neutralino*—a perfect WIMP candidate. We don't need supersymmetry to produce WIMPs, however; a simple extension of the Standard Model provides another candidate. Spontaneous symmetry breaking happens when the Higgs field rolls off the hump in the Mexican hat potential, picking out a particular field "direction," just as a pencil falls in a certain direction when it falls over.

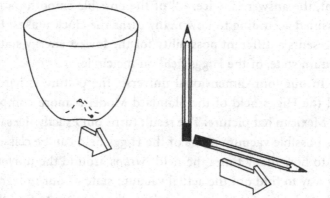

Let's indicate the final position of the Higgs field (or the pencil) by an arrow, as in the preceding figure. Now, remember that there is a potential like this *at every point in space*. There is the possibility, then, that the Higgs field will choose a different direction at different points in space. Instead of trying to picture how this happens in our universe, let's simplify things. Think of a circle with a lot of pencils balanced on end on top of it. They might all fall in the same direction, or they might fall in different directions. The Higgs field, though, must change slowly from one point in space to a nearby point, so let's require that the pencils' directions change smoothly, too. For instance, they might all fall outward, or they might form a pattern like the one on the right:

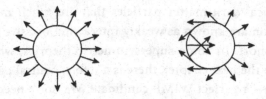

Think of the arrow's (or the pencil's) direction as a clock. How many complete turns does the clock make as we travel around the big circle on the right? If the pencils all fell in the same direction (say, they all fell to the right), the answer is zero; for the pattern on the left in the illustration (falling outward), the answer is once; and for the pattern on the right, the answer is twice. All of the possible smooth patterns can be classified according to how many turns the clock makes. Each pattern represents a different possibility for the lowest energy state, called the vacuum state, of the Higgs field on the circle.

In our four-dimensional universe, the picture is harder to draw, and the Higgs field of the Standard Model is more complicated than the Mexican hat picture. The result turns out exactly the same, though: The possible vacuum states of the Higgs field can be classified according to how many times the field "wraps around the universe." Is there any way to find out the actual vacuum state of our universe? There is, and we don't need to circumnavigate the universe to find the answer. It turns out that any vacuum state other than the simplest no-wrap, all-fall-the-same-direction solution violates a symmetry known as *CP symmetry*. CP symmetry involves the combination of mirror symmetry, or parity (that's the P), and an exchange of particles and antiparticles (that's the C). Experimentally, CP is only violated by a small amount. So our vacuum must be very nearly the no-wrap vacuum.

Out of an infinite number of possible vacua, why are we so close to the no-wrap one? That it "just happens" that way is as unlikely as hitting the bull's eye when throwing a dart blindfolded. However, it is easy to make a small modification to the Standard Model by adding a piece to the Lagrangian that gives all the wraparound vacua higher energy than the no-wrap vacuum.

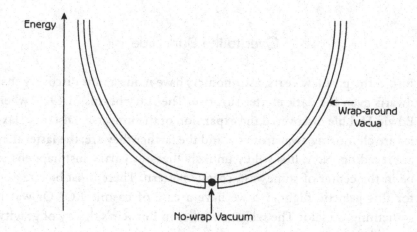

The diagram shows that the new term creates a "trough." From field theory we know that whenever we have an energy trough like this, the field can oscillate about the bottom of the trough. Field oscillations are just another way of describing particles, so this modification of the Standard Model implies that a new type of particle can exist. This (so far hypothetical) particle is given the name *axion*. Axions have the right properties to be WIMPs: zero electric charge and minimal interactions with other matter. Unfortunately, the theory gives few hints about the axion's mass. If these particles are abundant in the galaxy, they should have noticeable effects on star formation and the properties of supernova explosions. So far, all attempts to detect those effects have failed.

It is, of course, immensely embarrassing for physicists to admit that we have no idea what 85 percent of the matter in the universe is made of. At the same time, it is terribly exciting. Whether dark matter turns out to consist of neutralinos, axions, or something that has not yet been thought of, or whether the solution requires even more radical changes to basic physics, dark matter will certainly be a crucial part of physics beyond the Standard Model.

Over to the Dark Side

Just in the past few years, astronomers have made a new discovery that dwarfs even the dark matter surprise. The story begins in 1929, when Edwin Hubble discovered the expansion of the universe: Distant galaxies are all moving away from us, and the farther they are, the faster they are receding. Now, it is wildly unlikely that our galaxy just happens to be in the center of some vast, cosmic pattern. There must be a reason for this galactic flight. Do we have a case of cosmic BO? Or was it something we said? The real answer lies in Einstein's theory of gravity, general relativity.

General relativity posits a connection between the curvature of spacetime and the energy, including the mass-energy ($E = mc^2$), in the universe. Simply put, the equations say that *Curvature = Energy*. Einstein invented these equations to describe, for instance, planets orbiting the sun, or the gravitational attraction between galaxies. He soon began to wonder, though, whether the same equations could describe the spacetime structure of the whole universe. At this point, he made what he would later call his greatest blunder. Hubble had not yet discovered the flight of the galaxies, and so Einstein looked for a static solution to his equations, a solution that would describe an eternal, unchanging universe. In order to find such a solution, Einstein modified his original equation by adding a new term, which came to be known as the cosmological constant. This was a sort of fudge factor, unrelated to any of the known properties of gravity, which allowed a static solution to the (modified) equations of general relativity, a solution describing a universe that neither grew nor shrank but was balanced just on the edge. The cosmological constant represented a new kind of energy, inherent in space itself, evenly spread throughout the whole universe.

With Hubble's discovery, Einstein realized his mistake. The galactic recession Hubble described had a natural interpretation in Einstein's theory. It wasn't that all the galaxies were running away; it was that *the space between the galaxies was expanding.* To picture this expansion, take a partially inflated balloon and draw some dots on it with a marker. If you now continue to blow up the balloon, you will see that the dots move farther apart. Now think of the dots as galaxies. No matter which dot you are living in, you see all the other dots moving away from you. Moreover, the further away a dot is from your dot, the faster it moves. Space, in Einstein's theory, is like the rubber of the balloon: It can stretch. If all space is stretching all the time, then more distant galaxies will recede faster than closer objects, just as Hubble found.

The discovery of this cosmic expansion made Einstein's cosmological constant unnecessary. Physicists building mathematical models of the universe tended to ignore the cosmological constant. The original general relativity equations were perfectly capable of describing an expanding universe that started with a big bang. As we have seen, these models made testable predictions about the universe (for instance, the relative amounts of hydrogen and helium in the universe) that were in excellent agreement with observations. These physicists found that in all the models without a cosmological constant, an expanding universe would slow down. The universe might go on expanding forever at an ever-slowing rate, or the expansion might slow down enough that the universe began to recollapse. Astronomers began trying to measure the expansion rate more accurately to determine by how much the expansion was slowing. On the answer hinged the fate of the universe. An expansion that went on forever meant a universe that grew colder and colder as galaxies moved farther and farther apart. Stars would eventually run out of nuclear fuel, lose their heat, and stop shining. The lights would go out, and the universe would settle into an eternal, expanding night. On the other hand, a universe that recollapsed would begin to heat up again as galaxies were compressed closer and closer together. The temperature would rise, the density would rise, and the collapse of spacetime would accelerate until the universe disappeared in a big crunch, the opposite of a big bang. What was to be the ultimate fate of our universe, fire or ice?

It was only in 1998 that a new generation of telescopes (among them the aptly-named Hubble Space Telescope) became accurate enough to answer the question. By looking at supernovas in distant galaxies, astronomers were at last able to measure how the expansion rate was changing over time. The result was a complete surprise: Far from slowing down, the expansion was accelerating. To explain the accelerating expansion, astrophysicists returned to Einstein's "blunder." The cosmological constant supplies a uniform energy density at all points in space. This energy has no interactions with matter (other than gravitational), so it can't be seen; it is dark energy. Because of the way it enters into Einstein's equations, this dark energy counteracts the effect of normal mass and energy, causing the universal expansion to accelerate.

Whence does the dark energy arise? In Einstein's original model, it was simply another constant of nature, which, like the speed of light or Planck's constant, had to be determined by experiment. Particle physicists had a different suggestion. Recall Schwinger's description of a quantum field as a collection of harmonic oscillators, one at each point in space. Empty space, the vacuum, as physicists call it, is when all the oscillators are in their lowest energy state. We know, however, that *a harmonic oscillator has some energy even in its lowest energy state*. This vacuum energy exists at every point in space, and it has exactly the right properties for dark energy. That is, any relativistic quantum field theory predicts that empty space will be filled with dark energy.

It seems, then, that relativistic quantum field theory solves the mystery of dark energy. However, a problem arises when you try to calculate how much vacuum energy there is. Embarrassingly, relativistic quantum field theory predicts an *infinite* amount of dark energy. As it has no effect on particle interactions, the dark energy had always been ignored by particle physicists. After all, they were used to subtracting infinities as part of the renormalization technique. In the context of cosmology, though, the dark energy rises to critical importance. Is there a way to avoid the embarrassment of infinite energy and explain the dark energy at the same time?

Unfortunately, the answer seems to be no. We can get a finite value for the dark energy if we assume that new physics takes over at some energy scale, the Planck energy, for instance. This procedure results in a value for dark energy that is far too large, however. According to the supernova observations that implied its existence, the actual amount of dark energy contained in a thimbleful of empty space is equivalent to the mass of a few electrons. According to relativistic quantum field theory, though, the amount of dark energy in that same thimble of empty space is equivalent to a mass greater than that of all the known galaxies. Clearly, something is wrong with this picture. Why is the dark energy so nearly zero, but not exactly zero? The dark energy puzzle is one of the greatest unsolved problems of physics today.

The Muon: In Need of a Spin Doctor?

One of the great early successes of relativistic quantum field theory was the calculation of the electron's magnetic strength, what physicists call the magnetic moment. As we saw in Chapter 6, it is a measure of how fast the electron's spin axis will precess in a magnetic field. According to the Dirac equation, the magnetic moment should be precisely 2, but the cloud of virtual particles surrounding the electron alters the prediction slightly. The experimental measurement agrees with the predicted value to an astonishing one part in a billion.

Turning to the electron's heavier brother, the muon, things get considerably more interesting. The magnetic moment of the muon is evaluated in the same manner as for the electron: the basic Dirac equation prediction is still 2, and the cloud of virtual particles make a similar small change that is calculated via Feynman diagrams. Out to seven decimal places, the experimental and theoretical values are in agreement. However, in 2002, experimenters working at Brookhaven National Laboratory reported a new measurement, accurate to 10 decimal places. This result differs from the theoretical prediction only slightly, but the difference is more than twice the combined uncertainty in the experimental and theoretical values. This makes it unlikely, though not impossible, that the difference is due to chance.

In calculating the effects of the cloud of virtual particles surrounding the muon, we need to include not just the effects of virtual photons and virtual electron-positron pairs, but also virtual quarks, virtual Higgs particles, and, in fact, all the particles of the Standard Model. The same is true of the corresponding calculation for the electron. How is it possible, then, that the predicted electron magnetic moment is so accurate, while the muon prediction is slightly off? Here's one possibility: Suppose that in the real world there are some heavy particles that are not included in the Standard Model. These new particles would show up as virtual particles in the clouds surrounding both the electron and muon. It turns out, though, that because of the larger muon mass, any such heavy particles would affect the muon magnetic moment more than the electron magnetic moment. In other words, the small discrepancy in the muon's magnetic moment reported by the Brookhaven team could be evidence of new physics beyond the Standard Model. New experiments planned for Brookhaven and the Japan Accelerator Research Complex (J-PARC) may give results 10 times more precise than current values.

Glueballs, Pentaquarks, and All That

It is much more difficult to test QCD than it is to test other parts of the Standard Model because QCD deals with quarks and gluons, particles that always appear in combinations of two or more. It's simply impossible, as far as we know, to produce a beam of single quarks the way we produce a beam of electrons or neutrinos. Still, as we've seen, it's possible to test QCD in high-energy scattering experiments where the quarks behave like nearly-free particles.

In principle, QCD should also be able to describe the characteristics of bound states: how three quarks interact to form a proton, for example, and why it's impossible to remove and isolate a single quark. In a general sense, physicists think they know why these things happen, namely, that the long-distance behavior of the color force is just the opposite of its short-distance high-energy behavior: The color force grows stronger as you try to pull quarks apart. However, as the color force grows in strength, the approximations used to derive this behavior break down (this is discussed at greater length in Appendix B). The Feynman diagram expansion is no longer an accurate guide to the behavior of the color force. So, although everyone believes that QCD provides a correct theory of proton structure and of quark confinement, no one has ever been able to prove it.

In the face of these difficulties, some physicists have turned to computers, rather than pure mathematical analysis, to provide testable predictions. Even with computers, the problem at first seems intractable. Remember that we have to add together all of the possible paths that each particle can take—an infinity of paths, in fact. To make the problem tractable, physicists make a rather drastic simplification: They model spacetime as a finite grid, or lattice, of points rather than a continuum. The quarks and gluons are only allowed to move from one lattice point to another. This approximation, together with a host of calculational techniques for implementing it, goes by the name lattice QCD.

Even after such a simplification, the calculations require immense computing power, and lattice QCD physicists have for many years been some of the primary customers for the fastest supercomputers around. In recent years, some lattice QCD groups have even designed and built their own special-purpose computers, like the GF11, which uses 566 processors running simultaneously. Designed by Don Weingarten of IBM, the GF11 ran continuously for a year and a half to calculate the masses of a handful of the lightest mesons and baryons. The results agreed with the measured masses to within 5 percent. Obviously, we are far from the realm of one part in a billion accuracy. Five percent, though, was better accuracy than anyone had achieved for such a calculation before.

Calculations like Weingarten's are *retrodictions*: explanations after the fact of properties that are already known from experiments. More impressive would be an actual *prediction* of a previously unsuspected particle. All of the particles of the famed Eightfold Way turned out to be built out of either two or three quarks. Is it possible to build a particle out of four quarks, or five, or more? According to some theoretical considerations, the answer is yes. A *pentaquark*, for instance, might be built from two up quarks, two down quarks, and an antistrange quark. In 2003, Japanese experimenters at the Spring-8 accelerator near Osaka reported finding a bump in their data that could be interpreted as evidence of a pentaquark. Since then, nine other experiments have reported similar results. Still, researchers are not sure whether these experiments have actually succeeded in producing pentaquarks. More recent experiments, some of them with much better statistics, have failed to see any evidence for the pentaquark. There is disagreement, too, even among the experiments reporting positive results, about the mass and lifetime of the particle.

On the theoretical side, the situation is similarly fuzzy. Turning to pure QCD is hopeless: No one knows how to derive the properties of a proton, much less of a pentaquark. The original calculations that predicted the pentaquark used an approximate theory that captures some aspects of QCD. Later, lattice QCD calculations lent support to the prediction. Now, however, several different lattice QCD groups have carried out the calculation and predict different properties for the pentaquark, or no pentaquark at all.

An experiment at the Thomas Jefferson National Accelerator Facility (JLab) reported its results in April 2005. Because an earlier JLab experiment with a similar design had reported positive results, the new run was considered a crucial test of the pentaquark idea. After observing many more collisions than the earlier Jlab experiment, the new run turned up no evidence for the pentaquark. Yet another Jlab experiment is expected to report results sometime in 2005. If this search, too, turns up nothing, it may signal the demise of the pentaquark. Then theorists will have to address a different question: Why *don't* quarks combine in numbers larger than three?

Another exotic possibility is a particle called a *glueball*, built entirely from gluons. A glueball has no real quarks, though of course it has the virtual quark-antiquark pairs that are always present in QCD. The existence of glueballs cannot be derived directly from QCD; the complexity of the color force gets in the way, just as for particles made of quarks. As with pentaquarks, glueballs were first predicted using approximate versions of QCD, and lattice QCD calculations later confirmed the idea. With glueballs, though, the experimental situation is much less controversial. An experiment at Brookhaven National Laboratory way back in 1982 was the first to report evidence of a possible glueball. More recent experiments, like the L3 experiment at CERN and the ZEUS experiment at Germany's DESY accelerator, have confirmed the existence of the particle found at Brookhaven and have added several other glueball candidates. These particles certainly exist; however, no one can say for sure whether they are glueballs. On the one hand, there are the unusual uncertainties about the reliability of approximate methods and lattice QCD for the real world. On the other hand, it is not so easy to distinguish a glueball state from a bound state of a quark and an antiquark. A better theoretical understanding is needed to interpret these intriguing new particles.

An experiment now running at Cornell University called CLEO-c is examining the decay of the J/psi particle for evidence of glueballs. Experimenters hope to produce a billion J/psi particles from which a few thousand glueball candidates will be culled. This should be enough to nail down the rate at which the purported glueballs decay, and into which particles.

The discovery of glueballs would have profound implications for our understanding of QCD. Quarks were originally invented by Gell-Mann and Zweig in order to explain the patterns of spin and charge seen in subatomic particles. Later, the apparatus of gluons and the color force was added to explain why the quarks didn't just fly apart. As we have seen, there are good reasons to believe that quarks exist—the parton behavior in high-energy scattering experiments, for instance. There is no such direct evidence for gluons, however. There is, for example, no way to scatter off of a virtual gluon inside a proton. Glue-balls, if they can be proven to exist, will provide the first direct evidence of the existence of gluons.

In the Beginning, There Was Soup

Imagine running the universe backward in time. With the expansion of the universe in reverse, the temperature and density rise as we get closer and closer to the big bang. At about 1/10th of a second after the big bang, the temperature everywhere in the universe is now so high that atoms can no longer exist; they are torn apart into individual protons, neutrons, and electrons. Back up a little more to around 10 microseconds after the big bang and the neutrons and protons are shoulder-to-shoulder throughout the universe, together with a host of other particles that have been created by pair production from the extremely high-energy photons that now populate the universe. Each neutron or proton, we know, is like a little bag with three quarks inside. Go back in time a bit further and the boundaries between the neutrons begin to disappear, the "bags" merging like droplets of mercury running together. This point is known as the quark-hadron transition. With the boundaries between the neutrons and protons gone, the universe is now filled with a thick soup of quarks and gluons, which physicists call the quark-gluon plasma. Understanding this state of matter is crucial to understanding the first microseconds of the universe's existence. During the past decade, physicists have been trying to re-create this state in the laboratory.

The basic technique is to strip all of the electrons off a heavy atom, such as gold, accelerate it to high energy, and collide it with another heavy atom. This is like smashing two drops of water together so hard that they vaporize. Thousands of particles condense out of the mist and go flying off in all directions. Beginning with several experiments at CERN in the mid-1990s, and continuing with the Relativistic Heavy Ion Collider (RHIC) at Brookhaven National Laboratory in the past five years, experimenters have sought indications of the expected quark-gluon plasma. To date, these experiments have had some remarkable successes: they have achieved the extremely high temperature and density at which the quark-gluon plasma is expected, and they have found strong indications that matter behaves very differently under these conditions than in the usual single-particles collisions. All is not as expected, however, so the experimenters have been reluctant to say for certain that they have produced a quark-gluon plasma.

To mention just one predicted effect of the quark-gluon plasma, there should be fewer J/psi particles produced than in a comparable two-particle collision. The J/psi is a bound state of a charm quark and an anticharm quark. This quark-antiquark pair is created by pair production. Normally, the color force takes over at this point, binding the pair together fleetingly until one or both of the quarks decays. If the charm-anticharm pair is produced in the midst of a quark-gluon plasma, however, both particles will immediately attract a swarm of particles from the plasma. Just as a swarm of moths partially blocks the lamplight that attracts them, the swarm of plasma particles partially blocks the color force between the two quarks. As a result, the quarks are less likely to bind together into a J/psi particle.

As with pentaquarks and glueballs, the difficulty of calculating in QCD makes the interpretation of these experiments tricky. Suppression of the J/psi, for example, could be caused by something other than the plasma state. Some theorists expect a state different from the plasma state to form. One suggestion is that quarks will remain tightly bound to each other by the color force, even at the high temperatures reached in these collisions, and as a result will behave like "sticky molasses."[2] Research continues at RHIC, and a new experimental stage will begin in 2007 when the new CERN accelerator, the Large Hadron Collider (LHC), starts operation. The LHC will accelerate and collide beams of heavy nuclei, such as lead, reaching an energy density perhaps three times higher than RHIC's. A new detector named ALICE is being built to get a better look at the indications of quark-gluon plasma that have been seen in previous experiments. ALICE will also be able to compare production of particles like the J/psi with production of the Z^0, which is not expected to be suppressed since the Z^0 doesn't interact via the color force. These studies will give a much clearer picture of the properties of matter at the kind of temperature and density that existed in the first microseconds of the universe's existence.

Pentaquarks, glueballs, and the quark-gluon plasma have driven home an important point about the Standard Model: The problem of bound states of quarks and gluons is by far the least understood part of the theory. At the root of the matter is the issue of color confinement. We claimed, back in Chapter 8, that particles in nature only appear in color-neutral combinations. This claim, if it is true, explains why free quarks have never been detected and why gluons don't create a long-range force, as would be expected of a massless particle. The problem with this explanation is that it can't be derived from the theory of QCD itself—it must be imposed as an extra condition. The importance of color confinement has been recognized by the Clay Mathematics Institute of Cambridge, Massachusetts, which has offered a reward of $1 million for the solution of a related mathematical problem, the so-called "mass gap problem" of Yang-Mills theories. This was chosen as one of the seven Millennium Problems in mathematics, announced in Paris on May 24, 2000. Often, when testing the Standard Model, it is a case of theoretical predictions, sometimes decades old, waiting for new accelerators capable of performing the necessary tests. QCD is the exception to the rule: There are now exciting new experimental results just waiting for a better theoretical understanding.

Looking for the Higgs

As far as the Standard Model is concerned, the Higgs particle is the holy grail of high-energy experiments. It gives mass to the other particles and is the linchpin of spontaneous symmetry breaking. The Standard Model doesn't predict the Higgs mass directly, as it does the W and Z^0 masses. This has left experimenters to conduct a painstaking search at ever-higher energy in the hope of stumbling across it. In recent years, though, the situation has improved.

Earlier in this chapter, we saw that some measurable quantities, like the muon's magnetic moment, are affected by the presence of virtual particles, including, of course, virtual Higgs particles. By examining the processes that are most influenced by the virtual Higgs particles, we might expect to learn something about the Higgs mass. The procedure is complicated and the results uncertain, since there is still a lot of uncertainty in the values of many of the Standard Model parameters that enter the calculations. In particular, before the discovery of the top quark in 1995, there was too much uncertainty in its mass for any useful prediction of the Higgs mass to be made. In recent years, the top quark mass has been determined much more precisely; precisely enough that the effects of virtual Higgs particles can now be estimated. Careful measurements of the masses and decay rates of the W and Z^0 have led to a "best fit" Higgs mass of about 130 billion electron-volts.

The uncertainties are still large, but there is a 95 percent chance that the true mass is less than 280 billion electron-volts. CERN's LEP accelerator came tantalizingly close to the best fit value in 2000, before being shut down for the upgrade to LHC. Before the shutdown, the ALEPH experiment detected three events that could be interpreted as Higgs events. The experimenters petitioned for, and received, a one-month delay in the shutdown in order to look for more such events. They saw none, however, and today most physicists believe that the three candidate events were simply flukes. Certainly, they were insufficient to claim discovery of the Higgs.

As of 2005, Fermilab's Tevatron accelerator is the only machine with sufficient energy to have a chance of finding the Higgs. The Tevatron collides protons and antiprotons, reaching higher energies than LEP was capable of. The proton-antiproton collisions are, however, messier than LEP's electron-positron collisions, and it will take until about 2009 for experimenters to collect enough data to rule out, or rule in, a Higgs particle with mass up to 180 billion electron-volts. By that time, the LHC should be taking data, if all goes according to plan.

Experimenters can be confident that, if the Higgs behaves as the Standard Model predicts, there will be none of the uncertainty over its identity that plagues the pentaquark and glueball candidates. The Higgs must be a neutral, spin-zero particle with very specific, calculable interactions with the other particles. Experiments at LHC will try to measure as many of these characteristics as they can, but further confirmation may only come from the planned International Linear Collider (ILC).

Of course, the true symmetry-breaking mechanism that occurs in nature might look very different than the Standard Model Higgs particle. After all, the Standard Model Higgs was the bare minimum that needed to be added to the known particles in order to make spontaneous symmetry breaking work. Given all the awkwardness of the Standard Model, most physicists would probably be surprised, perhaps even appalled, if the Standard Model's Higgs turned out to be correct in every detail. The next chapter will reveal a plethora of alternative theories waiting in the wings, ready to step into the leading role when the expected misstep comes, each with its own predictions about what particle or particles should appear in place of the Higgs. Experimenters are acutely aware of these alternative possibilities, and must design their detectors accordingly. It is an immensely complex undertaking, but the reward is proportionate to the difficulty: Higgs physics may well reveal a deeper level of physical reality than any we have known so far.

Chapter 12

New Dimensions

What did we hope for, heating and reheating ourselves to absurd temperatures? As matter heats up it is subject to demonstrable change. Boiling in our vessel, our water molecules would begin to break down, stripping us back to elemental hydrogen and oxygen gases. Would this help us to see ourselves as we really are?

Heated further, our atomic structure would be ripped apart. He and she as plasma again, the most common state of matter in the universe. Would this bring us closer together?

At about a billion degrees K, give or take a furnace or two, he and she might begin to counterfeit the interior of a neutron star and could rapidly be heated further into sub-atomic particles. You be a quark and I'll be a lepton.

If we had the courage to cook ourselves to a quadrillion degrees, the splitting, the dividing, the ripping, the hurting, will be over. At this temperature, the weak force and the electromagnetic force are united. A little hotter, and the electroweak and the strong force move together as GUT symmetries appear.

And at last? When gravity and GUTs unite? Listen: one plays the lute and another the harp. The strings are vibrating and from the music of the spheres a perfect universe is formed. Lover and beloved pass into one another identified by sound.

—Jeanette Winterson, *GUT Symmetries*

For many years after the development of the Standard Model, all the experimental results pointed in one direction. From the W and Z^0 masses, to the parton picture of the proton, to the details of particle interactions and decay rates, everything seemed to indicate that the Standard Model gave an accurate picture of particle interactions. Theorists didn't wait around for experiments like those discussed in the previous chapter to reveal flaws in the Standard Model, though; they had reasons enough to suspect that the Standard Model wasn't the whole story. Some of these reasons were aesthetic: Why should a theory require 18 parameters, for example? Other reasons had to do with the shape of the periodic table of quarks and leptons. Why are there three families? Why are all the neutrinos left-handed? Is there any rhyme or reason to the masses of the particles? Why, for goodness's sake, is the top quark 45,000 times heavier than the up quark? The Standard Model explains none of this. Finally, of course, there is the omission of gravity from the theory. Physicists looking for answers to these questions have pursued two distinct paths. One path is to investigate theories with more *symmetry*. The other path is to look for deeper *structure*.

The Aesthetics of Particles, or Diamonds Are Not Forever

One of the ugliest things about the "repulsive" Standard Model is the way parity (mirror symmetry) is violated. As already mentioned, there are left-handed neutrinos but no right-handed ones. Electrons, though, have mass, and massive particles can't be purely left-handed or purely right-handed. How then to pair the electron and the neutrino, as required by the symmetry of the Standard Model? The problem is solved in a particularly crass manner. The left-handed part of the electron is paired with the neutrino, in an SU(2) doublet. The right-handed part of the electron remains all by itself—a singlet. This makes for an awkward arrangement, as if only one of a set of Siamese twins were married.

Much more aesthetically pleasing would be an underlying theory with left-right symmetry. To explain the left-handedness of the real world, this left-right symmetry must be broken. But we already know a way to make this happen—through spontaneous symmetry breaking. The idea is this: Start with a theory with left-right symmetry and introduce an "ultra-Higgs" particle in such a way that spontaneous symmetry breaking happens twice—first, the underlying left-right symmetric theory breaks down to the SU(3) × SU(2) × U(1) of the Standard Model, and then a second symmetry breaking happens, exactly as in the basic Standard Model. As we will see, theories that start out with left-right symmetry yield a bonus: They explain why neutrinos have a small but nonzero mass.

Another distasteful element of the Standard Model is the way color symmetry is tacked onto the unified electroweak theory. It would be much prettier to have a single underlying symmetry group that somehow contains the entire symmetry group of the Standard Model. Spontaneous symmetry breaking explains why the weak force is so different from the electromagnetic force: the W and Z^0 have mass, making the force they carry weak and short ranged, and the photon doesn't, making the electromagnetic force long ranged. Why shouldn't a similar process also explain the difference between the strong force and the electroweak force? All that is needed is to find the right underlying symmetry and the right pattern of symmetry breaking. Models that work this way are known as grand unified theories, or GUTs, for short.

To see how one symmetry group can contain another symmetry group, consider again the symmetry group of a circle. The circle can be rotated by any amount without changing it. Now consider the symmetry group of the sphere. One symmetry is rotation about the vertical axis. If you compare what happens to the "equator" in this rotation with what happens to the circle, you will see that the two transformations are identical. But the sphere has another symmetry, rotation about the horizontal axis. This is a different symmetry from the first; there is no way to obtain this rotation by any combination of rotations of the first type.

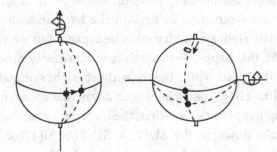

The symmetry group of the sphere is thus a larger group than the symmetry group of the circle, since there are two operations—that is, two distinct types of rotation that leave the sphere unchanged. But this larger group *contains* the symmetry group of the circle, in the sense that any of the possible rotations of the circle can be translated into a rotation of the sphere about one axis. In the same way, it is possible to find symmetry groups that are larger than the SU(3) x SU(2) x U(1) group of the Standard Model, and that contain it as a subgroup.

The smallest grand unification symmetry group that contains the Standard Model's symmetry group is called SU(5). The SU(5) GUT has been extensively studied, but is now ruled out by experiments. Let's consider a slightly larger symmetry group, SO(10) (pronounced "ess-oh-ten"). The periodic table of the fermions reveals three fermion families, each with an identical pattern of eight particles: one electron-like particle, one neutrino, and two flavors of quark, each flavor having three colors. The symmetry of the Standard Model accommodates this pattern, but it doesn't require it. The pattern suggests a more intimate connection between quarks and leptons. Could the electron-like particle and its neutrino be considered quarks of a different color? In that case, all the particles in each family should fit into a single multiplet. There are eight particles of the lightest family. The rules of SO(10) require that we count the right-handed and left-handed components separately. Since there are eight particles, there is a total of 16 components, *if* the neutrino comes in both left-handed and right-handed versions. The SO(10) symmetry group has a 16-component multiplet, just right to fit all the known particles of the lightest family, plus a right-handed neutrino. When the SO(10) GUT was invented, the presence of a right-handed neutrino was problematic: Neutrinos were believed to exist only in the left-handed version. As we saw in the previous chapter, though, recent experiments measuring the neutrinos coming from the sun have begun to convince physicists that neutrinos have a small but nonzero mass. Here's where the SO(10) GUT comes into its glory. When spontaneous symmetry breaking occurs, the right-handed neutrinos become extremely massive, so they no longer count as unwanted "light" neutrinos. (Oddly, the masses of the right-handed and left-handed neutrinos need not be the same.) Additionally, a mathematical result known as the seesaw mechanism guarantees that if the right-handed neutrinos are very heavy, then the left-handed neutrinos will be very light. This makes small (left-handed) neutrino masses a natural result of the theory rather than something imposed for no reason by the theorist. Any particle with mass *must* have both left-handed and right-handed versions. If the solar neutrino experiments are right, then the right-handed neutrino required by SO(10) symmetry must actually exist in nature. What was a drawback of SO(10) symmetry has become an advantage.

The SO(10) symmetry solves a host of other puzzles. First, it explains the family structure of the fermion table. Each family is a single SO(10) 16-component multiplet. This means that we can't add quarks to the model unless we add an electron-like particle and a neutrino at the same time. The symmetry *requires* that there are as many electron-neutrino families as there are quark families, in contrast to the Standard Model, where there was no such requirement from symmetry. Next, recall the odd asymmetry between left-handed and right-handed particles in the Standard Model. In the SO(10) model, left-handed and right-handed particles enter in a perfectly symmetrical way. The difference we observe in the real world between left-handed and right-handed particles arises from spontaneous symmetry breaking. The model is set up so that symmetry breaks in two stages. We introduce an ultra-Higgs particle, with its own Mexican hat potential, as well as the regular Higgs particle from the Standard Model. The parameters of the ultra-Higgs are chosen so that the left-right symmetry breaks first, collapsing the SO(10) symmetry to the SU(3) × SU(2) × U(1) of the Standard Model. The left-handedness of the universe arises as naturally as a tree falling, rather than being imposed by fiat. When the tree is standing, no one can tell which way it will fall. With the ultra-Higgs, similarly, there is no way to tell in the early universe whether the left-handed neutrinos will be light and the right-handed ones heavy or the other way around. It just happens that, in our universe, that was the way things fell out. In the second stage, SU(2) × U(1) symmetry breaks just as in the Standard Model.

Finally, SO(10) symmetry explains something that was completely mysterious in the Standard Model: why the electron and the proton have equal (but opposite) electric charge. The equality is astonishingly precise. Since everyday objects are made of something like 10^{24} atoms, even an extremely small difference in the two charges would be easy to detect. Experiments show that the difference must amount to less than one part in 10^{21}, in fact. This level of precision suggests the equality is the result of a deep underlying symmetry of nature. SO(10) provides just the symmetry needed. The way the Standard Model's symmetry fits into SO(10) symmetry requires that one quark flavor have exactly 2/3 of the electron's charge and that the other quark flavor have -1/3 of the electron's charge. Here is a fact of fundamental importance that was completely unexplained in the Standard Model, but is required by the SO(10) GUT.

The situation with particle masses is considerably more complicated. Remember that in the Standard Model it was the Higgs field that gave mass to the quarks and leptons. The Higgs couplings were not determined by symmetry, so these masses could be chosen to have whatever value we liked. In GUTs, the symmetry group relates some of the couplings, and so, in principle, gives numerical relations among the fermion masses. For instance, the electron and the down quark masses should be equal. Comparing with the experimental values—$m_e = 0.511$ million electron-volts and $m_d = 6.95$ million electron-volts—we see something has gone terribly wrong. These are not even close. However, two things need to be taken into account. First, the masses, like the coupling strengths in the Standard Model, are different when measured at different energy scales. The predicted relations clearly don't hold for the low energies accessible by experiment, but they might hold at the very high energies where SO(10) symmetry is unbroken. Second, the masses depend on the type of Higgs fields and the pattern of symmetry breaking. In GUTs, there are many possible choices for both. So there is still hope that we can find a Yang-Mills symmetry group, a set of Higgs fields, and a pattern of symmetry breaking that results in the correct fermion masses.

Here's how Howard Georgi describes his first encounter with GUTs:

> I first tried constructing something called the SO(10) model, because I happened to have experience building that kind of model. It's a group in ten dimensions. The model worked—everything fit neatly into it. I was very excited, and I sat down and had a glass of Scotch and thought about it for a while.
>
> Then I realized this SO(10) group had an SU(5) subgroup. So I tried to build a model based on SU(5) that had the same sort of properties. That model turned out to be very easy, too. So I got even more excited, and had another Scotch, and thought about it some more.
>
> And then I realized this made the proton, the basic building block of the atom, unstable. At that point I became very depressed and went to bed.[1]

Georgi's discouragement was premature. The crucial question: How *quickly* does the proton decay? If its lifetime is short, then matter as we know it could not exist. Long before the stars and galaxies formed, the hydrogen needed to make them would have disappeared. But if the proton's lifetime is long compared to the age of the universe, then SU(5) is still possible. Protons would stick around long enough for galaxies to form, for life to evolve, for civilization to arise. With a very long proton lifetime, we would have to be very lucky, or would have to work very hard, to see a proton decay at all.

Proton decay is a prediction of all GUTs, and it's easy to see why. Remember that electroweak unification gave us interactions involving the W that led to beta decay. GUTs introduce new intermediate particles, labeled X in the diagram below, that couple quarks with positrons, and quarks with antiquarks:

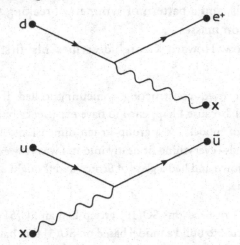

This means the proton can decay by the following process (among others):

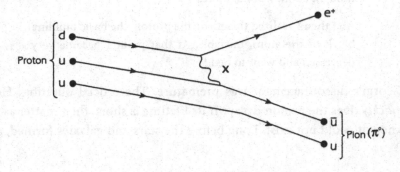

The proton decays into a positron plus a pion. The pion can then decay by matter-antimatter annihilation, and the positron can go find an electron and do the same. In the end, a proton has vanished and left only a few photons behind. If protons can vanish like this in a flash of light, then all matter is in danger of disappearing. Diamonds are *not* forever, neither is any other object, whether solid, liquid, or gas. Matter is unstable and, given enough time, will decay into electromagnetic radiation. How much time is enough? Georgi's paper on SU(5), titled "Unity of All Elementary-Particle Forces" (written with Sheldon Glashow), didn't give an answer, it only mentioned the possibility. The universe has been around for about 10 billion (10^{10}) years. If the SU(5) model predicted a proton lifetime much less than that, the model could be thrown out, as it would be incompatible with the observed fact that protons are still around. On the other hand, a lifetime like 10^{100} years would be impossible to measure in any conceivable experiment. A later paper by Georgi, Helen Quinn, and Steve Weinberg came up with a very exciting answer: 10^{31} years. This was safely above both the 10^{10} years since the big bang and the known minimum value (10^{27} years) of the proton lifetime, but it was low enough that it could be tested experimentally. It would require a strange new type of particle detector, however.

How can anyone hope to detect a decaying proton if the proton lives sextillion times longer than the current age of the universe? Obviously, it is out of the question to watch a single proton until it decays. Remember, though, that the "lifetime" of a particle is only its *average* time before decay, and according to the rules of quantum mechanics, the particle is equally likely to decay at any time. That is, protons formed at the big bang do not wait 10^{31} years and then decay all at once. According to the rules of quantum mechanics, decays occur at random, regardless of when the decaying particles were "born." Some will decay immediately, others only much later, so that the *average* time is 10^{31} years. This means that if we gather a large enough number of protons together, we have a high probability of seeing decay (if indeed they decay at all).

Proton decay experiments all work on this principle: Put a lot of stuff in one place and watch carefully to see if anything happens. A thousand tons of matter contain about 10^{33} protons, so, according to the SU(5) model, about 100 protons should decay every year. To avoid as much as possible the confusing effects of cosmic rays, experimenters go deep underground: to the Kolar gold mine in India, the Morton salt mine near Cleveland, Ohio, the Kamioka mine in western Japan. The IMB experiment in the Morton mine, for example, is a six-story high cube filled with purified water whose walls are lined with photon detectors called photomultiplier tubes. A proton at rest should decay into a pion and a positron moving in opposite directions. The pion then decays rapidly into a pair of photons. The positron that is emitted is moving nearly at the empty-space speed of light. The speed of light in water, however, is considerably less than the speed of light in empty space. Although, according to special relativity, nothing can travel faster than the empty-space speed of light, there is no law against going faster than the speed of light in water. Just like a jet airplane that produces a shock wave (a sonic boom) when it travels faster than the speed of sound, the positron traveling faster than the speed of light in water produces a cone of light, a "boom" known as Cerenkov radiation. Eventually, the positron encounters an electron from one of the water molecules and annihilates, producing another pair of photons. Thus, experimenters have a clear signal to look for: a ring of Cerenkov light and two pairs of photons with known energies.

In spite of some initial positive reports, physicists now agree that no genuine proton decays have been recorded in almost 20 years of experiments. The failure to see decays has pushed the minimum proton lifetime to about 7×10^{33} years. This rules out the SU(5) GUT model, but the SO (10) model remains possible.

Why Is There Something Instead of Nothing?

Pause for a moment and think about how far we have come. Modern physics began in the early twentieth century with the attempt to understand the spectrum of light emitted from hydrogen gas. At that time, the question of whether the universe had an origin, or whether there was fundamental mirror symmetry, would have belonged to the realm of religion or philosophy, not to physics. Thanks to general relativity and the big bang model, we now know the answer to the first question (yes, the universe had a beginning); and thanks to the Standard Model, we know the answer to the second (no, the mirror world is not identical to our world). GUTs allow us to probe other deep questions. One such question: Is matter eternal? We will know the answer if we ever detect proton decay. Here's another deep question: Why is there any matter in the universe?

To understand this last question, recall that every particle has an antiparticle with precisely opposite properties. When a particle and its antiparticle meet, they can annihilate each other, leaving only photons. The same process in reverse is what we called pair production: a photon is converted into a particle-antiparticle pair. Now, if the laws of physics are completely symmetric with respect to particles and antiparticles, then we would expect that the tremendous energy available in the early stages of the big bang would be transformed into any particle and its antiparticle in equal numbers. Why do we see today an overwhelming predominance of one sort of matter, the kind we call normal matter? Where is all the antimatter? Might some of the galaxies we see be made entirely of antimatter? This would resolve the matter-antimatter asymmetry, but there would necessarily be a region of intergalactic space where stray bits of matter met stray antimatter. The resulting annihilations would cause a faint but detectable glow. Antimatter galaxies cannot be the solution.

According to astrophysicists, it is just possible that distant clusters of galaxies separated from ours by immense amounts of empty space could be made of antimatter, as there would not be enough exchange of stray matter to cause a detectable glow. There's no plausible explanation, however, for such a separation of normal matter and antimatter. Therefore, physicists believe that all of the galaxies in the observable universe consist of normal matter.

Put the question another way: If particles and antiparticles were created in equal numbers in the early universe, why did they not all meet and annihilate shortly thereafter, leaving a universe full of photons and nearly devoid of matter? Russian physicist Andrei Sakharov considered the question in a 1967 paper and concluded that three conditions had to be met to end up with more matter than antimatter:

- The universe had to be out of equilibrium.

- Proton-decay-type processes had to be possible (what physicists call baryon nonconservation).

- These processes had to violate CP symmetry.

The rapidly expanding universe of the big bang model provides the nonequilibrium environment required by the first condition. We already know that GUTs satisfy the second condition. Do they satisfy the third?

As we've seen, the P of CP refers to the parity operation—the mirror symmetry previously discussed. We know that this symmetry is violated in the Standard Model. C refers to *charge conjugation*: exchanging all particles with their antiparticles. This symmetry is also violated in the Standard Model. Under C, a left-handed neutrino transforms into a left-handed antineutrino, a particle that doesn't exist in the Standard Model. The CP operation performs both actions at once: replace every particle with its antiparticle and reflect everything in a mirror. For example, the CP operation transforms a single neutrino (which are all left-handed, remember) into a right-handed antineutrino, a particle that *does* exist in the Standard Model.

Now, suppose we have a GUT with a hypothetical X particle that causes a proton-decay-type process. Let's say a down quark decays into an X particle and a positron (top, in the following figure). By rotating the diagram, we see that this same interaction would allow the X to decay into a positron and an antidown quark (middle). And by exchanging particles and antiparticles, we see that the anti-X decays into an electron and a down quark (bottom).

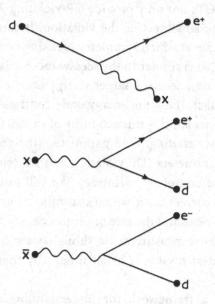

CP symmetry would guarantee that the decay of X and $\overline{X}$ happen at the same rate. If at some time in the very early universe a matter-anti-matter symmetry was present, with equal numbers of Xs and $\overline{X}$s, then CP symmetry ensures that the later universe will contain an equal number of down and antidown quarks. Only if CP is violated can an asymmetry arise from an initially symmetric situation. That is, if CP symmetry is violated, the decay of the X happens more often than the decay of the $\overline{X}$, or vice versa.

The Standard Model allows for, but does not require, CP violation; the amount of violation must be determined experimentally. This is fortunate; indeed, it is necessary, since a 1964 experiment by J. W. Cronin, V. Fitch, and two colleagues showed that CP is violated in the decay of kaons, particles composed of a down quark and an anti-strange quark. (Cronin and Fitch were awarded the 1980 Nobel Prize for this discovery.) The Standard Model does not, however, include any proton-decay-type processes, so even with CP violation it is utterly unable to explain the existing matter-antimatter asymmetry.

Remarkably, GUTs not only provide CP-violating X particles, they also can explain the smallness of the violation. It turns out that the larger the energy scale at which symmetry breaking occurs, the smaller the CP violation. This is similar to the seesaw mechanism that explains small neutrino masses, where a larger energy scale caused a smaller neutrino mass. In fact, the *same* energy scale controls both CP violation and neutrino masses. It is this economy of explanation that makes grand unification so exciting: One parameter (the energy scale) can explain multiple parameters (CP violation and neutrino masses) that would otherwise be completely arbitrary. We will not know whether this explanation is correct until we measure the neutrino masses, the amount of CP violation, and the rate of proton decay. If some GUT can be found that fits those measurements, though, we will have an answer to one of the greatest mysteries of all time: why there is something instead of nothing.

GUTs provide a framework for understanding a great deal of physics beyond the Standard Model. Neutrino masses, proton decay, CP violation, and the matter-antimatter asymmetry of the universe can potentially be explained. The greater symmetry of GUTs reduces the number of free parameters. However, the possibility of different types of Higgs fields and of multiple symmetry-breaking energy scales adds new free parameters. Often, extensive fine-tuning of the model is necessary to obtain agreement with experiments. This goes against the spirit of the enterprise: We wanted a theory with fewer free parameters, not more. Whether or not GUTs are, strictly speaking, more economical theories (more physics explained using fewer parameters) than the

Standard Model, they are, at least, a conceptual advance. They leave unanswered some of the big questions, however. Although GUTs tell us that the number of quark families must be equal to the number of lepton (electron-neutrino) families, they give no hint why there should be precisely three families. And, as usual, gravity is left out. Grand unified theories, in fact, are neither all that grand nor all that unified. Clearly, they are still far from a Theory of Everything. But that indeed may be ultimately a strength rather than a weakness. By not attempting to explain *everything*, they may actually succeed in explaining *something*. Experiments already in progress or planned for the next few years may reveal whether the universe has GUTs. Meanwhile, theorists have been looking elsewhere for answers to the big questions.

Inside the Quark

As the Standard Model racked up successful predictions throughout the 1970s and 1980s, it was natural for theorists to look to the higher symmetries of GUTs for further progress. As we have seen, though, these theories are not without problems, and there is arbitrariness in the selection of a Yang-Mills symmetry group, the Higgs fields, and the symmetry breaking patterns. Perhaps symmetry is the wrong way to go. Might *structure* provide the answer? The pattern of the chemical elements in Mendeleev's periodic table was finally understood in terms of the *structure* of the atom. Elements are not truly fundamental, they are made of protons, neutrons, and electrons. The details of the interactions between these constituent particles (specifically, the quantum mechanics of electron orbitals), explain the existence of chemical families, the columns of the periodic table. Similarly, the geometric patterns of Gell-Mann's Eightfold Way are now understood in terms of the quark structure of the new particles. Is it possible that the leptons and quarks in our new periodic table are not truly fundamental either, but are made up of still smaller entities?

Hypothetical particles that make up the quarks and leptons are known as *preons*. Let's try to build such a model by arranging the lightest lepton family as follows:

u_r	u_b	u_g	v_e
d_r	d_b	d_g	e^-

Suppose we think of the electron and its neutrino as another sort of quark: a fourth color, which we'll call lilac (or lepton; either way the abbreviation is L). Next, think of the particles in the top row of the table as having up-ness, while those in the bottom row have down-ness. Then, we can explain the form of the table by positing that each particle is composed of two preons. Let's invent one set of preons to carry the color property. There are four of these preons; let's call them R, B, G, and L, corresponding to the colors red, blue, green, and lilac. A second set of preons carries the up-ness or down-ness: call them U and D. Finally, we can build all of the particles in the table by supposing that the preons of the first set bind to the preons of the second set. For instance, a red up quark would be formed from the combination RU, whereas the electron would be LD. In addition, we need to invent a force to bind the two types of particles together. The hypothetical force is termed hypercolor, and it is taken to be a Yang-Mills-type force, just like the QCD color force.

Preon theories have to contend with the outlandish success of the Standard Model. Remember that the Standard Model predicts the electron magnet moment to an accuracy of one part in a billion. If the composite structure of the electron is not to mess up this prediction, then the binding energy due to the hypercolor force must be around 10 trillion electron-volts. Recalling that energy and mass are interchangeable ($E = mc^2$ again), we would expect the electron to weigh much more than the half a million electron-volts it actually does weigh.

Preon models thus explain some of the patterns of our table of quarks and leptons, but they come up short on several accounts:

- *Fermion families.* Preon models can accommodate any number of families. Why are there only three in nature?

- *Matter versus antimatter.* GUTs naturally predict proton decay and CP violation, necessary ingredients for explaining the matter-antimatter imbalance in the universe. No one has yet figured out how to incorporate these ingredients into preon models.

- *Fermion masses.* As we have seen, preon models give us no help here.

Although it is tremendously appealing to the imagination to postulate a deeper level of structure to quarks and leptons, theorists have been unable to match preon models to reality. They have been abandoned now in favor of other approaches.

Fermions and Bosons Unite!

Look back at the table of fundamental particles in Chapter 10. Setting aside for the moment the as-yet-undetected Higgs particle, the particles in the Standard Model are either massive fermions (assuming, as seems increasingly likely, that all the neutrinos have at least a small mass) or intermediate force-carrying bosons. In this division, we detect a reflection of the old separation of matter and force. Quarks and leptons are the constituents of matter; the intermediate particles are the forces. This division might cause some uneasiness, however. After all, don't the W^+, W^-, and Z^0 all have mass? Shouldn't they count as matter, too? And aren't bosons and fermions treated alike in relativistic quantum field theory?

In that fertile decade for elementary particle theory, the 1970s, a bold suggestion was put forth: Perhaps there is a fundamental symmetry between bosons and fermions. Initially, investigation of this symmetry was little more than a Mount Everest because-it's-there phenomenon. Theorists noted that such a symmetry was mathematically possible and set out to investigate it. Soon, though, this supersymmetry (affectionately abbreviated as SUSY) was discovered to solve some problems of string theories, which we will encounter shortly. This synergy created tremendous interest in both types of theory.

It may seem absurd to suggest a symmetry between bosons and fermions. A symmetry is an operation that leaves the world unchanged. Surely, if we replaced every fermion by a boson (and vice versa), the world would not be unchanged. Atoms, for instance, are made out of fermions: quarks and electrons. If these were suddenly changed into bosons, we'd certainly notice. There is no Pauli exclusion principle for bosons, so there'd be no periodic table of the elements, and therefore no chemistry, no biology, no stars, planets, or people.

An analogy might help here. Suppose fermions are women and bosons are men. If we replaced all the men with women (and vice versa) we should be able to tell that something changed, right? (For one thing, there would suddenly be a lot of single fathers.) Well, remember that what's fundamental in relativistic quantum field theory is the interactions. A man and a woman can interact and produce a daughter. Supersymmetry, as applied to interactions, requires that if we switch all the genders we still get a possible process. Switching genders in the sentence above we get: "A woman and a man can interact and produce a son." Nothing wrong with that. So maybe at the level of interactions supersymmetry makes sense.

Supersymmetry requires that each particle have a superpartner that has all the same properties except for spin. If the particle is a boson, you get the name of its superpartner by adding the suffix -*ino*. So the photon has a photino, the Higgs a Higgsino, and the W and Z have a Wino (pronounced "weeno," not "wine-o") and a Zino (pronounced "zeeno"), respectively, for superpartners. If the particle is a fermion, you get the name of its superpartner by prefixing *s-* to the name. So quarks have squarks and leptons have sleptons (no, I am not making this up); for example, we get selectrons and sneutrinos.

Now, for the important question: Does supersymmetry exist in nature? The answer: Nope. Supersymmetry insists, for example, that there exists a boson with that same mass and charge as the electron. Such a particle would be easy to find, but has never been seen. We know, however, that some symmetries are spontaneously broken. Perhaps this happens with supersymmetry, too.

Starting with the Standard Model and adding the smallest possible set of fields, we get the minimal supersymmetric standard model (MSSM, for short). Spontaneous symmetry breaking has a lot to accomplish in this model; somehow, all of the superpartners of the known particles have to become massive enough to have escaped detection. Here, we hit the first snag: There is no completely satisfactory way to accomplish this in the MSSM. Let's ignore this problem. After all, the MSSM is probably (like the Standard Model) only an approximation to some more complete theory. Assuming that supersymmetry is broken *somehow*, we come to a very exciting result. As we've seen, the three coupling constants of the Standard Model *almost* meet at a very high energy. It is said that "almost" doesn't count, except in horseshoes and hand grenades. Whether "almost" counts in physics depends on the experimental uncertainty, indicated below for the α_3 curve (which has the largest uncertainty of the three curves) by the dashed lines. We see that "almost" doesn't cut it in this case.

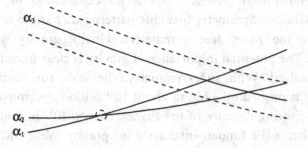

If all three intersection points fell inside the dashed lines, we could believe that they actually meet at a single point, and it is only the experimental inaccuracies that make it look like three different intersections. However, the third intersection point is well outside of the dashed lines, so we can't claim the three lines converge.

When we go to the MSSM, the picture changes. Remember the reason the coupling constants change with energy: because of the screening (or antiscreening) effects of the virtual particles. In the MSSM, there are more particles, so the constants change in different ways. Here, the intersection point is within the bounds of experimental uncertainty. If all three curves indeed meet at a single point, perhaps there really is a unification of the fundamental forces at that energy scale. This is the strongest hint to date of supersymmetry. Convergence of the couplings suggests that the MSSM may be an approximation to a more unified theory. Still, the MSSM has many of the same difficulties as the Standard Model. There are even more free parameters—the masses of all the superpartners, for example. There is no explanation of CP violation, baryon nonconservation, or neutrino masses. One possibility for a more unified theory is a supersymmetric grand unified theory. Supersymmetric GUTs have many of the advantages of normal GUTs: mechanisms for CP violation, neutrino masses that arise naturally, and proton decay. One general feature of these theories is that the proton, instead of decaying into pions (as in GUTs), decays into a kaon and a mu antineutrino. If experimenters ever observe proton decay, it will give a clear indication whether supersymmetry is realized in nature.

Unfortunately, supersymmetric GUTs also share the weaknesses of non-supersymmetric GUTs. There is no explanation for the three fermion families. Symmetry breaking patterns are hard to understand. There are too many free parameters. Finally, gravity is still not included. The potential importance of gravity is clear from the energy scales involved. With supersymmetry, the scale for unification of couplings is pushed up a bit, to about 10^{16} billion electron-volts. This is within spitting distance of the Planck energy, 10^{19} billion electron-volts, which is the fundamental scale for gravity. Well, OK, it's still a factor of a thousand too small, but, considering that the unification scale is more than a trillion times larger than the mass of the W (which gives the electroweak energy scale), a factor of a thousand is nothing to get excited about. Maybe unification will only be achieved when gravity is taken into account on an equal footing with the other forces.

String Music

The journey into GUTs and supersymmetry has opened up new horizons of scientific investigation. A physicist of 100 years ago would have laughed at the idea of explaining the origin of all matter. We do not yet have the answers to these new questions, but the fact that they can be given serious consideration as *scientific*, rather than philosophical or religious, questions, reveals how far we have come. Other facts remain, though, that show us how far we still have to go.

The energy scale of grand unification is not very far from the fundamental energy scale for gravity, yet none of our theories has anything to say about gravity. In fact, the accepted theory of gravity, general relativity, is in fundamental conflict with quantum mechanics. In our discussion of quantum mechanics, we learned that a particle can be in a superposition state, where it is equally likely to be found on either side of a barrier. According to general relativity, a particle that has mass causes a dimple in spacetime. Where should the dimple be for the particle in the superposition state? There can't be a half-dimple on the left and a half-dimple on the right—that would correspond to *two* particles, each with half the mass. The dimple can't be entirely on the left—that would mean we could discover the particle's location by measuring the gravitational effects. If we *know* the particle is on the left, there is no longer 50 percent probability of it being on the right. We have destroyed the superposition state. The same argument tells us the dimple can't be on the right, either. The only solution is to allow spacetime itself to be in a superposition state. That means we need a quantum mechanical theory of gravity.

String theory is the first theory to bring quantum mechanics and general relativity together. The fundamental premise is very simple: Instead of describing fundamental point-like (zero-dimensional) particles, the theory postulates a world consisting of one-dimensional strings.

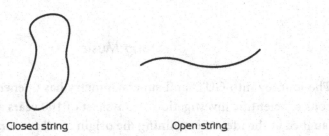

Closed string   Open string

The natural length of these strings is postulated to be the Planck length, obtained in the same way we found the Planck energy: by combining the fundamental constants c (the speed of light), G (the gravitational constant), and $\hbar$ (Planck's constant of quantum mechanics):

$$\text{Planck length} = \sqrt{\frac{Gh}{c^3}} = 10^{-35} \text{ meter}$$

This length is as much smaller than a proton as a badger is smaller than our galaxy. Such a tiny string would, for all practical purposes, behave like a point particle. Well then, why not stick to particle theories? The hope is that string theories give you more bang for the buck: More physics (all of the particles and gravity) is explained using fewer parameters. There are only five basic types of string theory (as compared to an infinity of GUTs), and each one has only one free parameter, the string tension. To encompass all of physics with a single number—that would truly be a Theory of Everything.

Strings give us a natural explanation for the bizarre processes of relativistic quantum field theory. The basic interaction is when a string splits into two strings. Slicing up this diagram (known as the pair of pants diagram) shows us the process in detail.

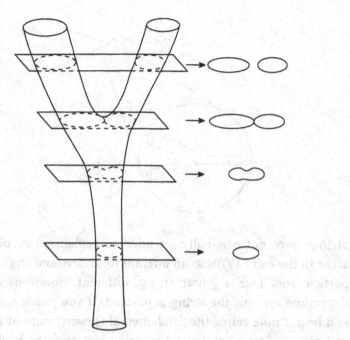

The basic interaction is simply a single string pinching itself and separating into two strings. This interaction does not have to be added into the theory the way we added particle interactions to produce a relativistic quantum field theory. The interaction is already implicit in the description of a single string. In relativistic quantum field theory, we had to add all the ways that a particle could go from one point to another. In string theory, we need to add all the ways a string can go from one place to another. This includes the kind of pinching that happens in the pair of pants diagram. Since this interaction is automatically part of string theory, there are no new parameters involved. Because of the incredibly small scale of strings, we will see the process as a particle decaying into two particles:

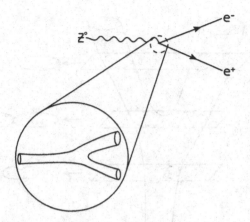

Strings were not originally intended to explain all of physics. They arose in the early 1970s as an attempt to understand the ever-growing particle zoo. Like a guitar string, different vibrations are possible depending on how the string is plucked. If you pluck a guitar string, you hear a note called the fundamental (lowest) note of the string. If you place a finger lightly on the string just over the twelfth fret and pluck the string again, you hear a note that is an octave higher. Doing the same on the fifth fret gives you a note two octaves above the fundamental. Similarly, string theory strings can vibrate in many ways. Each mode of vibration corresponds to a particle of similar properties but different mass, since more energy is needed to produce the higher modes. This early version of string theory was only partially successful in explaining the pattern of particle masses. After the Standard Model became the accepted theory of elementary particles, strings were abandoned by all but a few tenacious researchers.

In 1974, two of these researchers, Joel Scherk and John Schwarz, realized that string theory predicted a particle with spin 2. It had long been known that the graviton, the hypothetical particle that carries the gravitational force just as the photon carries the electromagnetic force, should have spin 2. Scherk and Schwarz realized that string theory, reinterpreted as a theory at the Planck scale, could potentially be the first-ever quantum theory of gravity. There were, however, two serious problems with the quantum version of this theory. First, it predicted the existence of a particle that traveled faster than the speed of light, called a tachyon. Second, the theory only worked if there were 26 space-time dimensions.

In the 1980s, string theory met supersymmetry. It was a match made in heaven. Their offspring, superstring theory, had no embarrassing tachyons. It necessarily included fermions, which were missing from the older string theory. Instead of 26 dimensions, quantum mechanics required 10. Ten dimensions, while an improvement over 26, might still seem excessive. After all, we only see four dimensions: three space dimensions and time. But theorists soon found ways to hide the extra six dimensions, a crucial matter for matching superstrings to the real world. At low energies, where the strings look like particles, these theories became identical to supersymmetric grand unified theories. You got different supersymmetric GUTs depending on how you went about hiding the extra dimensions. As we have seen, supersymmetric GUTs can encompass all the known elementary particles, so superstrings potentially explain all the physics of the Standard Model.

Most exciting of all was the discovery that strings *require* general relativity. The quantum version of string theory isn't mathematically consistent unless the equations of general relativity are satisfied. Einstein's beautiful theory of gravity could actually be *derived* from string theory. This stunning result gave strings a tremendous boost among the theoretical physics community.

String theories have a very different structure than particle theories, and theorists have had to develop new mathematical tools for dealing with them. The extreme complexity of the mathematics has kept strings from fulfilling many of the early hopes for them. There may be only one free parameter in the theory, but there are billions of different ways of hiding the six extra dimensions. How to choose the correct one? In principle, the theory itself should provide the answer, but in spite of 20 years of work on strings, no one yet knows how to find it.

In spite of these outstanding issues with string theory, it remains an active area of theoretical research. After all, when it comes to a theory that potentially unifies all of known physics, string theory is still the only game in town. And when physicists are investigating the extremes of energy, temperature, and density that existed in the fractions of a second after the big bang, a unified theory is not a luxury; it is indispensable.

The Standard Model lets us take an imaginary trip back in time to a microsecond after the big bang, a time when the universe was a hot soup of quarks, gluons, photons, leptons, Ws, and Zs. According to general relativity, as we run the film backward in time toward the big bang, the temperature and density keep increasing, past the GUT scale at which the symmetry between the strong and electroweak forces is restored, temperature and energy rising up and up until time zero, the instant of the big bang itself, where temperature and density are infinite and the entire universe is collapsed into a single point. This is obviously nonsense; general relativity is signaling its breakdown. Before string theory, physicists dealt with the problem of time zero by bypassing it. Simply assume that at some very small time after the big bang, the universe was in a very dense, very hot state, then all of the predictions of the big bang model follow. The infinite temperature and density at time zero are just ignored.

Strings offer the tantalizing possibility of a glimpse of what happened at, or even before, time zero. Unfortunately, little can be said with any amount of certainty due to the extreme difficulty of performing string calculations. A few promising suggestions have been made, however.

To approach the issue of what happens at or before the big bang, let's begin with a different question: If string theories are 10-dimensional, why is our space-time apparently four-dimensional? One way of hiding the extra six dimensions is to assume that they are rolled up tightly into loops of the Planck length. A garden hose viewed from a distance appears one-dimensional, but when seen up close appears two-dimensional:

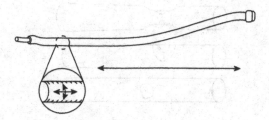

In a similar way, we can suppose that the six extra dimensions are very small, like the dimension around the hose's circumference. This still doesn't answer the question. Why shouldn't five, seven, or none of the dimensions be wrapped up small like this? Or all ten? Indeed, since the Planck length is the natural scale of strings, the most natural universe is one in which all of the dimensions are comparable to the Planck length. This leads us to phrase the question another way: Why are four of the dimensions so large?

When a dimension has a small circumference, strings can actually wrap around the circle:

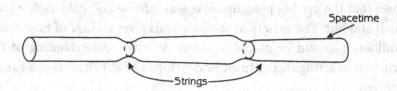

Like rubber bands around a rolled-up newspaper, these wrapped strings tend to keep the universe from expanding in that dimension. When strings collide, though, they can unwrap:

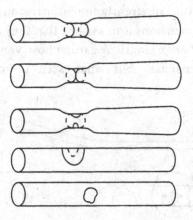

String theorists discovered that these collisions were likely to happen if (at most) three space dimensions were involved. As strings unwrap, these dimensions expand rapidly, rather like the inflationary model, leaving the remaining six space dimensions small. The resulting picture is of a universe that begins as a hot ball of nine space dimensions, the tenth dimension being time. The dimensions of this ball are all around the Planck length, the natural size for objects in string theory. Suddenly, three of the spatial dimensions begin to expand, an event that we interpret as the moment of the big bang.

If string theory is speculative, then scenarios such as the one just described are ultraspeculative. A different string-based scenario supposes that the pre-big bang universe was infinite and cold, rather than small and hot. The variety of the proposals gives an idea of how much confidence should be placed in them. A better understanding of the structure of string theory must be developed before these issues can be resolved.

Although string theory might reveal what the universe was like before the big bang, it can't tell us where the universe itself came from. Or can it?

Even before strings, physicist Stephen Hawking noticed that any theory that unites general relativity and quantum mechanics must treat spacetime as a field, the way it is treated in general relativity.

When this field is zero, there is no spacetime. There is a possibility, then, that the origin of the universe could be explained as a transition from a state with no spacetime to a state with a spacetime like ours: a real creation *ex nihilo*. Hawking, together with James Hartle, showed how an expanding spacetime like ours could be connected, at time zero, to a timeless space.

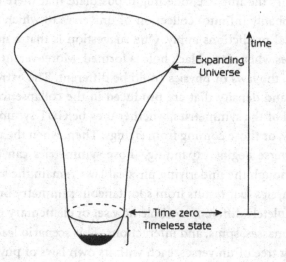

Time *begins* at time zero; there is no time on the other side of this line, so there is no such thing as "a time before the big bang." This scenario saves us from an unending series of "What happened before that?" at the price of trying to imagine how our universe of space and time could arise from (no, nothing can "arise" if there's no time), er, be conjoined with, a space with no time. (See what I mean?)

In any of these scenarios, a subtle philosophical problem arises when we take into account the nature of quantum mechanics. Quantum mechanics only gives the probability of a certain outcome. That is, in a collection of a large number of identically prepared systems, quantum mechanics tells us how many will have outcome A, how many will have outcome B, and so on. The difficulty comes in trying to apply quantum mechanical theories to our universe: We only have one universe. We can never assemble a large collection of identically prepared universes, so we can never test the quantum mechanical predictions of these theories. We have an impasse: Any theory that hopes to explain the big bang must unite general relativity with quantum mechanics, but any quantum mechanical theory of the universe is untestable since there is only one universe.

To skirt the impasse, one might postulate that there is in reality a large (probably infinite) collection of universes to which the quantum mechanical predictions apply. One suggestion is that a new baby universe arises whenever a black hole is formed. Moreover, in each of these universes, the laws of physics could be different. The extremes of temperature and density that are produced in the collapse to a black hole restore all of the symmetries, whether they be GUT symmetries, supersymmetry, or those coming from strings. Then, when the newly formed baby universe begins expanding, those symmetries can break in new ways. Although the underlying physical laws remain the same, the low-energy physics that results from spontaneous symmetry breaking might look completely different: a whole new set of elementary particles, with different masses, spins, and interactions. This scenario leads to an ever-branching tree of universes, each with its own laws of physics.

Occam's razor might suggest that postulating the existence of an infinite number of unobservable universes is a poor solution. Nor is that the only problem with this scenario. Testing the quantum mechanical predictions still seems impossible, since the properties of any universe other than our own can never be determined. However, universes with the largest number of black holes will be the most successful at generating new universes. In a sort of Darwinian manner, universes in which the physical laws are tuned to produce the maximum number of black holes should quickly dominate the tree. If we assume that our universe is one of these highly probable ones, then this scenario makes a definite prediction: Any small changes of the laws of physics should result in production of fewer black holes. At our current state of knowledge, this prediction is hard to test. For one thing, we don't know which fundamental parameters can be independently adjusted. In the Standard Model, for instance, the mass of the electron and the mass of the up quark are independent, but in GUTs, the two masses are related. A more fundamental problem is that our universe might not be one of the most probable ones. The most probable universe might be entirely unsuited for life—for example, it might be a Planck-sized ball that flashes into and out of existence, never growing large enough for stars, let alone life, to form. Perhaps the universe, a unique event, can only be understood in an entirely new, still to be formulated, framework.

A theory can't be considered scientific unless it makes predictions that can be tested in experiments. Over the 20 or so years of string theory's existence, several experimental tests of the theory have been proposed.

- Very massive particles formed by strings that wrap around the extra dimensions of spacetime could have been formed in the big bang. Some of these might be stable and exist even today. Searches for these particles have not been successful.

- In GUTs, the quarks are required to have 1/3 or 2/3 of the electric charge of the electron by the symmetry of the theory. In strings, the same result is achieved by the way strings wrap around the extra dimensions. Other wrappings would produce particles with 1/5, 1/7, or 1/11 of the electron's charge. Searches for fractionally charged particles have likewise been unsuccessful.

- Some versions of string theory predict that gravity will behave differently for very small objects than it does for the large objects for which we can measure it: things weighing a pound or so, up to planets, stars, and galaxies. So far, the most sensitive tests of gravity have not revealed any such deviations.

- String theories suggest that physics at tiny (Planck-length) distances is very different from what we normally see. Although any conceivable accelerator has far too little energy to probe such small scales, it is possible that the effects of string theory might show up as a cumulative effect when light travels over very long distances. For instance, light from distant galaxies must traverse billions of light-years of space, and so we might expect string physics to blur the images of those galaxies. However, no such blurring has ever been seen.

With so many negative experimental results, why do physicists continue to be excited about string theory? Normally, when a theory fails an experimental test, we expect the theory to be discarded. Note, however, the abundance of weasel-words in the preceding paragraph: "might," "could," "some." The truth is there is no one theory called "string theory." There are actually many string theories, all of which make different predictions. The experiments that have been performed rule out some versions of string theory, but there are literally billions of other versions that could still be true. The original hope, of a theory with fewer parameters than the Standard Model, has not been realized (so far, at least). String theory has been more successful as a quantum theory of gravity, and has significantly advanced our understanding of black holes. In this, it is almost the only game in town. Even if it is not clear which version, if any, of string theory might describe our universe, the theory provides a framework in which to ask questions that couldn't otherwise be asked.

At the moment, string theory remains a beautiful but unproven idea. Perhaps future theoretical advances will reveal how the Standard Model arises from the strange string world, or provide clear tests of the theory at accessible energies. Recent results have shown that the five basic versions of superstring theory are all interconnected, raising hopes that they form part of a still deeper theory, which has been labeled M-theory. Until these theories generate and pass experimental test, though, they will remain speculation.

In Search of the Theory of Everything

New discoveries, such as neutrino oscillations, dark matter, and dark energy, are forcing us to the conclusion that the Theory of Almost Everything has major deficiencies. Theoretical considerations, like the incompatibility of the quantum-mechanical Standard Model and the classical theory of general relativity, point in the same direction. But for 30 years, the Standard Model has been proving its worth. In experiment after experiment, it has provided a unified structure for understanding the behavior of elementary particles and unrivalled accuracy in its predictions. For the first time in the history of science, there is a single theory that provides an accurate mathematical description of matter in all its forms and interactions, gravity always excepted. The Standard Model is truly the crowning scientific accomplishment of the twentieth century.

As science progresses, new ways of describing the world are invented. If the new paradigm provides a more accurate description, it takes the place of the old. A scientific revolution occurs. In many cases, though, the older theory is not discarded completely, but retained as a useful approximate theory. Newton's theory of gravity, more than 300 years old, is still used for planning rocket launches, satellite orbits, and the paths of interplanetary space probes. Only rarely is it necessary to employ the more accurate description provided by general relativity.

Maxwell's electrodynamics equations have been superseded by the equations of QED, but Maxwell's continue to be used, except when dealing with very high energies or the world of the very small. Even if a more complete, more unified theory is discovered, the Standard Model will undoubtedly continue to be used as an accurate, if approximate, theory of particle interactions.

This is a thrilling time for fundamental physics. The greatest outstanding theoretical mystery, how to reconcile relativistic quantum field theory with general relativity, is beginning to yield to the assaults of the string theorists. New experimental results are starting to reveal how the Standard Model needs to be modified. There are great expectations for the new accelerator being built at CERN, the LHC. If it succeeds in producing the Higgs particle, much may be learned about whatever deeper theory lies beneath the Standard Model. If it fails, that failure will itself imply that the Standard Model is in need of major modifications.

Isaac Newton wrote in 1676, "If I have seen further it is by standing on the shoulders of giants." Never have we had so high a perch as we do now, never have we seen so far, and never have we had such hopes of new vistas to be discovered. From this height, we see an amazing panorama. Early in this century, scientists trying to understand the bright lines of color in the spectrum of their hydrogen vapor lamps invented the puzzling theory of quantum mechanics. Weaving together quantum mechanics and special relativity, relativistic quantum field theory depicted a world in which particles did the impossible—traveled faster than light, went on every path at once—yet this picture resulted in an accurate representation of the world around us. The Standard Model took this bizarre foundation, added a handful of observed particles and a few hypothetical ones, and became the most accurate and wide-ranging theory known to science. With this theory, we can peer inside the proton or journey back in time to the first millisecond of the universe's existence.

The (so far) speculative theories that go beyond the Standard Model promise to take us to even more bizarre unexplored realms. At the same time, they hint at ultimate limits to our knowledge. The fundamental length scale of string theories is the Planck length, 10^{-35} meters. Are strings made up of still smaller bits, described by an even more fundamental theory? To answer this question, we would need a way to probe even smaller distances than the Planck length. In particle theories, there is a wavelength associated to every particle. A particle will only work as a probe if its wavelength is at least as small as the features you want to see. As the particle's energy increases, its wavelength decreases, so you can see features as small as you like. All you need to do is give the particle enough energy.

Strings are different. For low energies, a string behaves like a particle, but if you give a string an energy larger than the Planck energy, the string grows larger. You can't see details smaller than the Planck length using a string that is larger than the Planck length. In other words, if string theory is true, it is impossible to investigate distances smaller than the Planck length. This is not a matter of insufficient technology, nor is it an indication of breakdown of the theory. No nonsensical results arise; indeed, true stringy behavior is just beginning to show up at the Planck energy. Rather than hinting at still deeper structure, string theory firmly declares that there is *no* deeper level that is experimentally accessible. There is an ultimate and insuperable limit to knowledge of the very small. So, it is argued, strings could really provide an ultimate theory, beyond which no further structure will ever be discovered.

Many theoretical physicists seem to assume that there is a unique Theory of Everything just waiting for us to discover it. Godlike, it summons the universe into being from nothing, creates all matter and energy, and directs every event. The history of physics tells a different story. Every theory has eventually been replaced by a new theory, often one that gives a radically different picture of reality, as quantum mechanics did in replacing Newtonian mechanics. Moreover, there is clearly an advantage in having different mental images for the same physics, as in the Feynman (particle) and Schwinger (field) images of relativistic quantum field theory. There is no doubt that physics is producing better and better approximations to reality. Can we really hope to achieve an understanding that represents reality exactly?

Suppose we had such a theory, or rather, a candidate. How would we know if it is the Theory of Everything? Our knowledge is limited. The universe, presumably, does not stop at the boundaries of our telescopes' vision. We can never know anything about the regions beyond. We have explored only a tiny portion of the energy range up to the Planck scale. Even if string theory were somehow verified through the entire range, we would still be in ignorance of what happens at energies a million, or a million million, times larger. There will always be realms of distance and energy that are beyond our reach. Finally, what explains the existence of the physical laws themselves? Even a Theory of Everything can't reveal why that theory, and not some other, is the one that describes our universe.

Ultimately, we may need to accept our own limitations. Perhaps all physical theories are approximations. Perhaps we need different, complementary, approaches for a unique event such as the big bang and other, repeatable, experiments. Limitations need not prevent us from pushing for a deeper understanding; indeed, knowing our limitations may help achieve that understanding.

As Frank Wilczek put it:

> So I expect that in ten to fifteen years we will know a lot more. Will we know everything? More likely, I think, is that as we learn many additional facts, we will also come to comprehend more clearly how much we don't know—and, let us hope, learn an appropriate humility.[2]

Quarks and the Eightfold Way

The quark model, invented by Gell-Mann and independently by Zweig, shed light on the earlier Eightfold Way classification of particles. The mysterious patterns of SU(3) symmetry were suddenly seen to be a result of the internal structure of subatomic particles.

The mesons, in the new scheme, are those particles consisting of one quark and one antiquark. The pions (π^+ and π^-), for instance, look like this:

		Charge	Isospin
π^+	(u)($\bar{\text{d}}$)	$\frac{2}{3} + \frac{1}{3} = +1$	$\frac{1}{2} + \frac{1}{2} = +1$
π^-	($\bar{\text{u}}$)(d)	$-\frac{2}{3} - \frac{1}{3} = -1$	$-\frac{1}{2} - \frac{1}{2} = -1$

What about the pi-zero (π^0)? There are actually *two* ways to combine the up and down quarks to make a particle with zero electric charge: either combine up with antiup, or down with antidown. Which of these is the pi-zero? The answer, oddly enough, is "both." Quantum mechanics tells us that it is possible to put a particle in a superposition state, where it has, say, a 50 percent probability of being behind Door Number 1 and a 50 percent probability of being behind Door Number 2. According to quantum field theory and the quark model, the pi-zero is a superposition of the two possibilities $u\bar{u}$ and $d\bar{d}$, with 50 percent probability of each. Resist the temptation to ask, "Well, which one is it *really*?": the quantum nature of the pi-zero doesn't allow that question to be answered. "Obviously," you object, "that's nonsense! The up quark has electric charge 2/3, while the down quark has charge −1/3. All we need to do to find out what the pi-zero is *really* made of is to measure the charges of the quarks inside it." Well, what do we find if we attempt this? The experiment has never been done, because the pi-zero doesn't last long enough to do it—the quark (whichever one it is) and the antiquark annihilate rapidly, releasing their energy in the form of photons. But, if we could do the experiment, the rules of quantum field theory tell us that half the time we would measure charges of +2/3 and −2/3 (corresponding to the up-antiup possibility) and half the time we would measure them as −1/3 and +1/3 (corresponding to down-antidown)—just like the house in Chapter 4 that was red on some days and blue on others. The reactions involving the pi-zero that we *can* measure completely support this surprising conclusion.

The property of strangeness, which seemed like an arbitrary invention, has a very natural interpretation in terms of quarks. A particle containing one strange quark has strangeness −1, a particle with two strange quarks has strangeness −2, a particle with one antistrange quark has strangeness +1, and so forth. This rule, together with the electric charge assignments of the quarks, allows us to disentangle the Eightfold Way multiplets in terms of their quark content. The lightest meson octet looks like this:

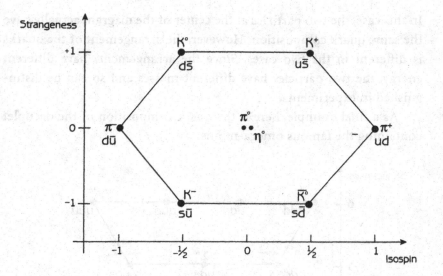

The two particles at the center of the diagram are the pi-zero, discussed already, and the eta-zero (η^0), which is a superposition of $u\bar{u}$, $d\bar{d}$, and $s\bar{s}$.

The lightest baryons also form an octet:

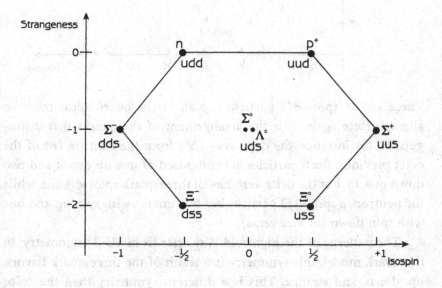

In this case, the two particles at the center of the diagram actually have the same quark composition. However, the arrangement of the quarks is different in the two cases. Since the arrangements have different energy, the two particles have different masses and so can be distinguished in experiments.

As a final example, here is the quark composition of the decuplet containing the famous omega-minus:

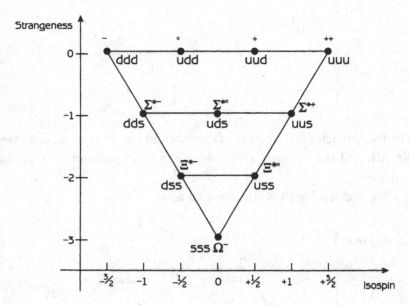

These are all spin −3/2 particles, so all of the quark spins must be aligned. Here again, it is the arrangement of the quarks that distinguishes, for instance, the delta-zero (Δ^0) from the neutron (n) of the octet previous. Both particles are composed of one up quark and two down quarks, but the delta-zero has all three quark spins aligned, while the neutron, a spin −1/2 particle, has two quarks with spin up and one with spin down (or vice versa).

The patterns of the Eightfold Way arise from SU(3) symmetry. In the quark model, this symmetry is a result of the three quark flavors: up, down, and strange. This is a different symmetry than the color

SU(3) symmetry of QCD, which is a result of the fact that each quark flavor exists in three different colors. Color symmetry is an exact symmetry, while the Eightfold Way is only approximate, since the three quark flavors have different masses. We now know that there are actually six quark flavors, not three. Clearly, the Eightfold Way will be of no assistance in classifying particles containing charm, bottom, or top quarks. We could contemplate an extended symmetry that includes all six quark flavors, but the wide range in mass of the quarks means such a symmetry is very inexact, and so of limited usefulness.

Appendix B

Asymptotic Freedom

Chapter 8 set out the claim that the color force decreases in strength as we go to higher energy and shorter distances, a property known as asymptotic freedom. This appendix will reveal how asymptotic freedom comes about. This will give me a chance to show you how physicists make use of Feynman diagrams. The color red is represented by a thick line, green by a medium line, and blue by a thin line.

The basic gluon-exchange interaction between two quarks looks like this:

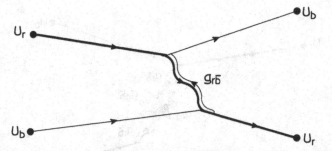

Let's now try to include the effects of virtual particles on this interaction, which is to say, include the Feynman diagrams involving interactions more complicated than simple gluon exchange. Remember, though, that the more interactions there are in a diagram, the smaller the contribution of that diagram to the final result. Physicists classify the Feynman diagrams by the number of loops that appear in them. To calculate the corrections to the basic gluon-exchange interaction, we need to include two one-loop diagrams. The first is this:

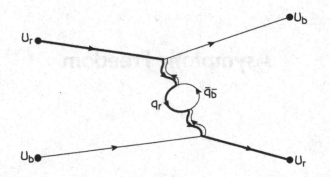

Here a red-antiblue gluon turns into a virtual quark-antiquark pair, denoted by $q\bar{q}$. The virtual quark-antiquark pair then annihilates to re-form the gluon before it hits the second quark. Notice that the quark-antiquark pair that gets created can be any flavor: up, down, strange, charm, top, or bottom. So there really should be six diagrams that look exactly the same except for the type of quark in the central loop. Also, notice that the color of the quarks in the loop is fixed: one must be red and the other must be antiblue.

The second process is this:

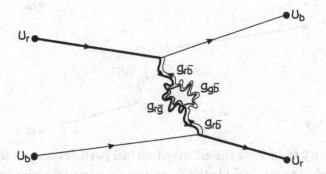

Here the exchanged gluon splits into two gluons, as allowed by the Feynman rules for the color force (see Chapter 8 and Appendix C). The two gluons then recombine before hitting the second quark. Notice that the color of the gluons in the loop is only partially determined by the colors of the incoming quarks. The left-hand gluon in the loop, for instance, could be red-antigreen, red-antiblue, or red-antired. So, there are really three diagrams like this one: one diagram for each color.

The first of these two processes is the same sort of thing that occurs in QED; for instance, when a photon splits into an electron-positron pair. This process is the origin of the screening effect in QED. Its effect in QCD is the same: it increases the strength of the quark-quark inter-action. The second process, though, has the opposite effect: it reduces the strength of the interaction. It is like a tug-of-war with the flavors on one team and the colors on the other. There are as many interac-tion-increasing diagrams as there are flavors, and as many interaction-reducing diagrams as there are colors. A rather long and messy calculation (the calculation that won Gross, Politzer, and Wilczek the Nobel Prize) reveals that each diagram of the second type contributes an amount five and a half times larger than a diagram of the first type. That is, each player on the color team can pull five and a half times harder than each player on the flavor team. Even though there are only three colors, the color team wins. Taken together, the whole set of dia-grams will have the effect of reducing the strength of the quark-quark interaction.

Now, recall that, as the collision energy gets higher, the more com-plicated Feynman diagrams become more important. That means that the reduction caused by the one-loop diagrams is greater at higher energy than it is at lower energy. This is exactly what we set out to prove: at higher energy the quark-quark interaction gets weaker. Physi-cists call this asymptotic freedom; the larger the energy of the electron beam used to probe the proton, the more the quarks inside the proton behave like free particles. This property is ultimately responsible for the scaling behavior observed in high-energy inelastic collisions.

For any particle, larger energy means larger momentum. The Heisenberg uncertainty principle ($\Delta x\, \Delta p > \hbar$) implies that larger momentum means shorter distance. That is, a high-energy electron beam probes the *short-distance* behavior of the color force. On the other hand, when we try to pull quarks apart we are probing the *long-distance* behavior of the color force. By the same reasoning as before, we should expect that at long distances the strength of the color force *increases*. Now, however, we run into a problem. The whole Feynman diagram method was based on the idea that the more complicated diagrams contribute less to the final result, and so can be ignored. This is only true as long as the interactions are small, though. If the interactions are large, then the more complicated diagrams could contribute as much as, or more than, the basic gluon exchange diagram. When that happens, all bets are off. The one-loop calculation is no longer reliable, even for an approximate answer. If we were able to include two-, three-, or four-loop diagrams, we still wouldn't know if we had done enough. Maybe the five-loop contribution is larger than any of those we have already calculated. There is just no way of knowing when it's safe to stop calculating.

To sum up the discussion, the one-loop approximation becomes increasingly accurate as we go to higher energy collisions and probe shorter distances. This means that asymptotic freedom is a rigorous result of the theory: the higher the energy, the more confidence we have in the approximation we are using. On the other hand, quark confinement is *not* a rigorous result of the theory. At longer distances, the Feynman diagram method fails. We know that the color force should get stronger for a while, but when it gets too strong the approximation no longer works. There's no guarantee that the force won't get weaker again at some even longer distance. Of course, we expect that it just keeps getting stronger, and this is why we've never seen a free quark. However, there's no known way to derive this result from the theory of QCD. As a result, the bound states of quarks and gluons remain a great mystery.

Interactions of the Standard Model

The Standard Model can be written either in mathematical form, using the Lagrangian function in Chapter 10, or in pictorial form, using Feynman diagrams. The two forms are completely equivalent if one knows the Feynman rules for translating between diagrams and mathematics. This appendix will display all of the Feynman diagrams for the Standard Model, thereby giving a complete list of all the fundamental interactions of the particles. These diagrams summarize the rules by which the universe runs. After giving all of the diagrams, I'll indicate how they relate to the mathematical formulation of the theory.

First, there are the propagators. The propagator gives the probability for a given particle to go from one place to another in the absence of any interactions. The mathematical form of the propagator depends on the particle's spin; this is indicated in the language of diagrams by using a straight line for a fermion, a squiggly line for a spin-one intermediate particle, and a dashed line for the spin-zero Higgs, as in the examples here:

$$e^- \bullet \!\!\longrightarrow\!\! \bullet \; e^-$$

$$\gamma \; \bullet\!\!\sim\!\!\sim\!\!\sim\!\!\sim\!\!\bullet \; \gamma$$

$$H \; \bullet\!-\!-\!-\!-\!\bullet \; H$$

There is a propagator for each particle in the Standard Model, of course. Since they all look alike, I won't bother to list them all separately. The labels on the ends specify which specific particle is involved.

Now, the interactions. All of the interactions of the gluons among themselves, and of the W, Z^0, and photon with each other, are determined by the Yang-Mills symmetry of the theory. For gluons, there are the three-gluon and four-gluon interactions from Chapter 8:

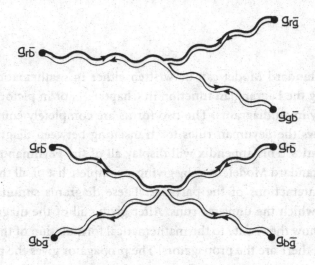

The electroweak intermediate particles interact among themselves as follows:

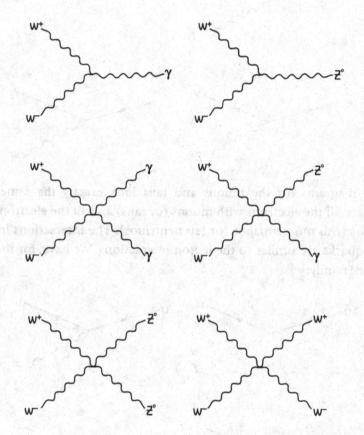

(Here, and everywhere in this appendix, I leave out diagrams that can be obtained from these by exchanging all particles with their antiparticles.)

Next, we need the interactions between the fermions and the intermediate particles. The diagrams involving electrons and neutrinos are these:

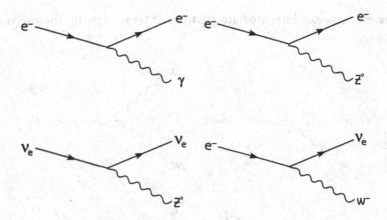

The diagrams for the muons and taus look exactly the same. Just replace all the electrons with muons (or taus) and all the electron neutrinos with mu neutrinos (or tau neutrinos). The interactions involving quarks are similar to the lepton interactions. We have, for the first quark family:

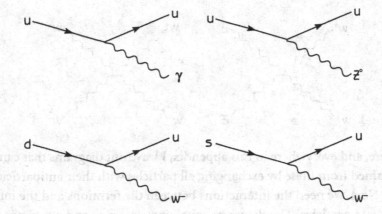

The last diagram actually involves a quark from the second family, the strange quark (s), as well as a quark from the first family, the up quark (u). This is an example of *quark mixing*, in which quarks of different families get jumbled up. The amount of mixing is controlled by a parameter known as the Cabibbo angle. It's one of the eighteen

Standard Model parameters mentioned in Chapter 10. For the other quark families, the Feynman diagrams look just the same as the first three diagrams above. There are additional quark mixing diagrams as well, along with appropriate additional parameters.

The weak interactions of the fermions described by the diagrams above give a deeper understanding of beta decay. In beta decay, a neutron decays into a proton plus an electron plus an antineutrino. We can now describe this process using the quark model of the neutron, together with the interactions of the Standard Model. The picture looks like this:

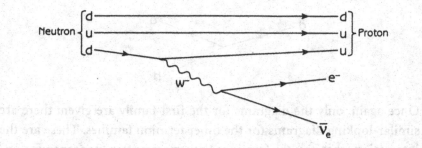

Here we see a neutron (*dud*) that emits a (virtual) W⁻, turning into a proton (*duu*). The W⁻ then decays into an electron (e⁻) and an antineutrino ($\overline{\nu}_e$)—exactly what's needed for beta decay.

The gregarious Higgs boson interacts with just about everybody. It interacts with the intermediate particles:

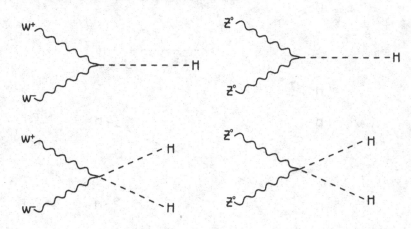

and with the fermions:

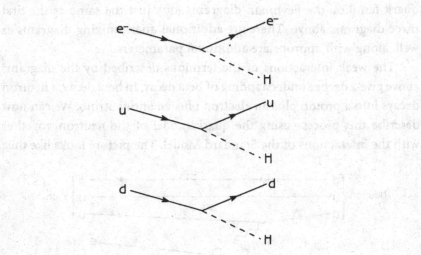

Once again, only the diagrams for the first family are given; there are similar-looking diagrams for the other fermion families. These are the interactions that give the fermions their masses after spontaneous symmetry breaking. Note that there is no interaction between the neutrinos and the Higgs; thus, the neutrinos remain massless in the Standard Model.

Finally, the Higgs interacts with itself:

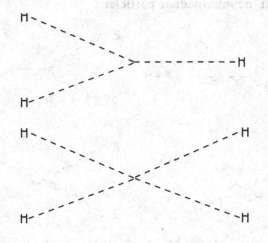

These self-interactions generate the Mexican hat potential so crucial for spontaneous symmetry breaking.

The interactions just listed reveal how the Higgs gives mass to the fermions. Picture the Mexican hat potential with its central hump. The Higgs field is zero at the center of the potential—at the top of the hump, that is. When symmetry breaks, the Higgs field rolls off the hump and ends up in the trough, where *it is no longer zero*. As we saw in Chapter 9, this shift of the Higgs field occurs at every point in the entire universe. So there is a background of Higgs field throughout the universe. Picture yourself as a fermion: wherever you go, there is a Higgs field. You cannot escape it. Like an annoying little brother constantly plucking at your sleeve, it's always there pulling on you, thanks to the interactions in the diagrams above. It is this constant tug of the background Higgs field that gives mass to the quarks and leptons.

Now you know everything that physicists have learned about the fundamental processes of the universe! Any (non-gravitational) interaction of the known fundamental particles involves some combination of the basic processes listed here. In order to generate actual numbers from the diagrams, though, we need a more mathematical description.

The mathematical description of the Standard Model starts from the Lagrangian function.

$$L = -\frac{1}{4}G^2 - \frac{1}{4}W^2 - \frac{1}{4}F^2 + (\nabla H)^2 + \mu^2 H^2 - \lambda H^4 + i\sum_j \bar{f}_j \nabla f_j - \sum_{jk} c_{jk} \bar{f}_j H f_k$$

As we learned in Chapter 10, G describes the gluon field and its interactions, W stands for the SU(2) field (describing the W^+, W^-, and Z^0 particles and their interactions with each other), F is the U(1) field, and H is the Higgs. The symbol Σ here means "add the following things together," while each f_j (for different values of the index j) stands for one of the fermions: the electron and its kin, the neutrinos, and the quarks. The last term in the Lagrangian therefore involves two fermion fields and the Higgs field. The c_{jk} are the coupling constants for the fermion-Higgs interactions. These numbers determine the masses of all the quarks and leptons. The actual numerical values of the coupling constants are not determined by the theory and must be learned from experiments. The symbol ∇ represents a mathematical operation called a "covariant derivative."

There is a direct relationship between each of the Feynman diagrams given above and some piece of the Lagrangian. Look, for instance, at the term $-\lambda H^4$. The symbol λ stands for a numerical parameter, while H^4 represents four Higgs fields multiplied together. In terms of Feynman diagrams, this is exactly the piece that gives us the four-Higgs interaction shown above. The parameter λ tells the likelihood of the four-Higgs process: it is the coupling constant that must be associated with that particular Feynman diagram. Since no one has ever detected a Higgs particle, let alone measured the probability for Higgs particles to scatter off each other, no one knows the numerical value of λ. It is one of the eighteen parameters that determine the Standard Model of which, as yet, we have no experimental knowledge.

Every term in the Lagrangian corresponds to a Feynman diagram in just the same way. Not all of the interactions are as easy to pick out as the four-Higgs interaction: some interactions are hidden by the notation. The interactions of the gluons among themselves are hidden inside the G^2 term, while the interactions between the fermions and the intermediate particles are hidden inside the ∇ symbol. The purpose of all this sleight of hand is to make it possible to write the Lagrangian in a very compact form. To go from the Lagrangian to the Feynman diagrams, we need to make all of the hidden terms explicit. Like a Swiss army knife, the Lagrangian is only useful when you pull out the bits that are folded up inside.

Naturally, neither the Lagrangian nor the Feynman diagrams given here are enough for you to start calculating probabilities for particle interactions. There are other details (most of them having to do with the spin of the different particles) that I have left out of this simplified Lagrangian. Even beyond such simplifications, there is the whole structure of relativistic quantum field theory, which tells how to take a Lagrangian, or a set of Feynman diagrams, and extract actual numerical predictions. If you want to learn all of the details, you'll just have to go to graduate school in physics!

Glossary

Alpha particle A helium atom with its electrons stripped off. An alpha particle consists of two protons and two neutrons.

Antimatter Particles with all of the properties of normal matter reversed. An antimatter particle can combine with its normal counterpart: The two annihilate each other and their energy is converted into photons.

Asymptotic freedom The strange property of QCD that tells us the strong force gets weaker when color-carrying particles are closer together.

Baryon Originally a term for any heavy particle, it is now used for any particle made up of three quarks (or three antiquarks), such as the proton and the neutron.

Beta decay A form of radioactivity in which an electron (formerly called a beta ray) is released. The simplest beta decay process is when a neutron decays into a proton, an electron, and an antineutrino. Beta decay is caused by the weak force.

Boson Any particle with a whole-number spin value—for example, the Higgs particle (spin 0) or the photon (spin 1).

Cathode In a cathode-ray tube (CRT), such as a TV tube, the cathode is the source of the electrons that eventually strike the screen to make the picture.

CERN (European Organization for Nuclear Research) The world's largest particle accelerator facility in Geneva, Switzerland, where the W and Z^0 particles were discovered.

Charm One of the six quark flavors. Particles containing at least one charm or anticharm quark are said to be "charmed."

Classical Descriptive of the physics that predates quantum theory, or for which there is not yet a quantum description (as for gravity).

Cloud chamber An early form of particle detector in which particles are seen by their trails through some form of vapor.

Color The property of quarks that makes them subject to the strong force. The three colors (usually called red, blue, and green) are related by SU(3) symmetry. In nature, only color-neutral combinations such as (red + blue + green) or (red + antired) seem to exist.

Confinement (of quarks) The curious fact that quarks are never found floating about by themselves; they are always confined inside protons, neutrons, or other particles. Quark confinement was given a theoretical foundation with the discovery of asymptotic freedom by David Gross, Frank Wilczek, and David Politzer, who shared the 2004 Nobel Prize in Physics for the discovery. Confinement still lacks a complete theoretical explanation, though.

Conserved quantity Take any of the basic measurable physical quantities (like mass, charge, spin, position, and velocity) and combine them mathematically according to some fixed formula. If the resulting combination does not change when the particles interact, it is a conserved quantity. Examples are the total energy and total momentum of a system.

Constructive / destructive interference Combine waves from two (or more) different sources, and you will find that the waves reinforce each other at some locations, causing larger ups and downs (constructive inference), and cancel each other at other locations, causing reduced oscillation or none at all (destructive interference).

Coupling constant A number that expresses the strength of a particular interaction.

CP symmetry / CP violation The C (charge conjugation) operation exchanges all the particles of a theory with their antiparticles. The P (parity) operation reflects everything into its mirror image. If a theory remains unchanged when these two operations are performed simultaneously, it has CP symmetry. The observed asymmetry between matter and antimatter can only be explained if CP symmetry is violated.

Dark energy The generic term used for whatever is driving the accelerating expansion of the universe.

Dark matter Astronomical observations of galaxies and galaxy clusters reveal that they contain much more mass than can be ascribed to the visible matter (stars and dust) they contain. Dark matter is the term for whatever accounts for the remaining mass. This missing mass cannot be normal matter or antimatter.

Decay Atoms or particles that spontaneously transform into other particles are said to decay. The decays are random, but they can be characterized by the *lifetime*, the average time it takes for a collection of the particles to decay.

Dirac equation The equation, formulated by P. A. M. Dirac, that gives a relativistic description of electron and positron behavior.

Eightfold Way A classification of particles by SU(3) symmetry. The Eightfold Way was superseded by the quark model.

Electric charge A property of some particles that causes them to attract or repel one another. All known particles are either uncharged (neutral) or have charge equal to a whole number multiple of one-third of the proton's charge. In this book, electric charges are always given as a multiple of the proton charge.

Electric field The means by which a charged particle influences other charged particles. Classically, the electric field is represented by an arrow at every point in space. In relativistic quantum field theory, the electric field is the result of a cloud of uncountable virtual photons.

Electromagnetic waves Mutually reinforcing changes in the electric and magnetic fields that carry energy from one place to another. Radio waves, microwaves, X rays, and visible light are all electromagnetic waves.

Electroweak force In the Standard Model, the electromagnetic force is unified with the weak force into a single force, known as the electroweak force.

Electron Along with the proton and neutron, one of the particles that make up ordinary atoms. The electron's electric charge is equal but opposite to the proton's, and the electron's mass is about 1/2000 of the proton's mass.

Electron-volt The amount of energy gained by an electron moving through one volt of electric potential. Also used as a unit of mass through $E = mc^2$.

Elementary particle A particle that cannot be separated into constituent particles.

Energy A conserved quantity that measures the capability of an object to influence other objects. Energy comes in two forms: kinetic energy that an object has due to its motion, and potential energy that an object has due to its interactions with other objects.

Family The elementary fermions of the Standard Model can be organized into three families. The first (or lightest) family consists of the electron, the electron neutrino, the up quark, and the down quark. The second family consists of the muon, mu neutrino, charm quark, and strange quark, all of which have charge and spin identical to the corresponding first-family particles. The third family consists of the tau, tau neutrino, top quark, and bottom quark, again with the same pattern of charges and spins.

Fermilab (Fermi National Accelerator Laboratory) One of the largest particle accelerator facilities in the world, located in Batavia, Illinois. Fermilab experiments discovered the bottom and top quarks and the tau neutrino.

Fermion Any particle with spin 1/2, 1 1/2, 2 1/2, and so on. Fermions obey the Pauli exclusion principle: Two identical fermions can never occupy the same quantum state.

Fine structure The details of the line spectrum of an element.

Fine structure constant The coupling constant that expresses the strength of the electromagnetic force. Denoted by α, it is related to the electron charge e, Planck's constant $\hbar$, and the speed of light c by

$$\alpha = \frac{e^2}{\hbar c}$$

Flavor The name for the type of quark: up, down, charm, strange, top, and bottom. Each quark flavor has a different mass. The term does not, of course, imply that quarks can be distinguished by tasting them.

Frame of reference A system that can be used to locate the positions of objects and the time of events.

Fundamental units It is much easier to talk about particle properties in terms of fundamental units rather than laboratory units. In this book, electric charge is given in multiples (or fractions, in the case of quarks) of the proton's charge, and spin is measured in units of Planck's constant, $\hbar$. For instance, using fundamental units we can say, "the electron is a particle with charge -1 and spin 1/2," rather than, "the electron is a particle with charge -1.609×10^{-19} coulombs and spin 5.28×10^{-35} Joule-seconds."

Gamma ray A high-energy photon, generated in some radioactive decays and in particle-antiparticle annihilations, as well as in other processes.

General relativity Albert Einstein's theory of gravity as a consequence of curved spacetime.

Gluon The massless intermediate particle that carries the color force.

GUTs (grand unified theories) Theories in which the Standard Model's SU(3) × SU(2) × U(1) symmetry is replaced by a larger symmetry. These theories reproduce all of the predictions of the Standard

Model, but they predict additional particles and interactions (such as proton decay) that do not occur in the Standard Model. GUTs do not unify gravity with the other forces.

Hadron A generic term for any particle built of quarks.

Harmonic oscillator Any vibrating system in which the vibration frequency does not depend on the amplitude. Almost any vibration can be treated as approximately harmonic for small oscillations.

Higgs particle The symmetry-breaking particle of the Standard Model. It has not yet been detected in experiments, but there is hope that it will be found in the next generation of CERN experiments.

Hole In Dirac's electron theory, a hole is a positive charge that is left behind when an electron is lifted out of the negative-energy "sea." In modern terms, we simply call it a positron.

Intermediate particle A particle whose exchange is responsible for one of the forces of the Standard Model: strong, weak, or electromagnetic. Intermediate particles all have spin 1. They are also known as Yang-Mills bosons or gauge bosons.

Invariance If a theory predicts that the result of any experiment is independent of some property (for instance, the location or orientation of the experiment), we say the theory has an invariance.

Isospin A particle property invented to explain certain types of nuclear interactions. In the Standard Model, these interactions are explained in terms of quarks.

Kaon (K^+, K^-, K^0, and $\overline{K}^0$) A meson formed of one up or down quark and one antistrange quark, or else of one strange quark and one antiup or antidown quark. Kaons were the first strange particles discovered.

Lagrangian The mathematical expression that summarizes all of the properties and interactions of the particles in a relativistic quantum field theory.

Lambda-zero particle (Λ^0) A baryon formed of an up quark, a down quark, and a strange quark.

Least action principle The global formulation of the laws of physics, which states that a particle will follow the path along which the sum of its Lagrangian values is the smallest.

Lepton Any of the lightweight particles: electron, muon, tau, or their neutrinos. The leptons are all fermions and have spin 1/2.

Lifetime (of a particle) See Decay.

Line spectrum The pattern of bright lines that forms when the light emitted by excited atoms or molecules is passed through a prism.

Magnetic field The field created by a moving electric charge. Elementary particles such as the electron also have an intrinsic magnetic field that cannot be ascribed to the motion of charge since the electron, as far as we know, has no structure.

Magnetic moment A measure of the magnetic strength of a particle (or of a magnet). It can be measured by applying a magnetic field and measuring the rate at which the particle's spin axis rotates.

Mass Roughly speaking, the mass of an object is simply its weight. More precisely, the mass is the object's inertia—that is to say, its resistance to a force.

Mass-energy Thanks to Einstein's $E = mc^2$, we know that matter can be converted into energy and vice versa. It therefore makes sense to lump mass and energy together in a single term.

Maxwell's equations The equations that describe the behavior of classical electric and magnetic fields.

Meson Originally, a term for the middleweight particles: heavier than the electron but lighter than the proton. In modern usage, any particle formed from one quark and one antiquark.

Mexican hat potential The particular form of the Higgs field's self-interaction that allows spontaneous symmetry breaking to occur. Drawn on a graph, it looks like a sombrero.

Muon (μ^+, μ^-) A lepton with the same electric charge and spin as the electron, but a mass 200 times larger.

Neutrino (ν_e, ν_μ, ν_τ) A lepton with no electric charge and a mass much smaller than the electron's mass. Neutrinos come in three flavors: electron neutrino, muon neutrino, and tau neutrino. They only interact via the weak force.

Neutron A neutral, spin $-1/2$ particle that is one of the three main constituents of ordinary matter. The neutron is slightly heavier than the proton. It consists of one up quark and two down quarks.

Noether's theorem States that for any continuous symmetry of a theory there is a corresponding conserved quantity.

Nucleon In isospin theory the neutron and the proton are considered two different states of a single particle, the nucleon.

Nucleus The heavy central core of an atom, consisting of protons and neutrons and bound together by the strong force.

Omega-minus (Ω^-) A baryon formed of three strange quarks.

Pair production The process in which a particle and its antiparticle are produced from pure energy.

Particle accelerator A device for accelerating particles, usually electrons or protons, to extremely high energies in order to perform collision experiments. Accelerator designs include cyclotrons, synchrotrons, and linear accelerators. Today's most powerful accelerators are many miles long.

Parity The operation of flipping an object into its mirror image. For instance, if you raise your right hand, your mirror image raises its left hand. If the parity operation leaves a situation or theory unchanged, it is said to be *invariant under parity*.

Parton The name used for a particle that is a constituent of a proton or neutron. *Parton* is used instead of *quark* when the full theory of QCD is not assumed.

Pauli exclusion principle The principle that two identical fermions cannot occupy the same quantum state.

Perturbation expansion Solving a problem by starting with only the largest influences, then adding successive corrections. In relativistic quantum field theory, the procedure of adding successively more complicated Feynman diagrams is a perturbation expansion.

Photoelectric effect The flow of electric current produced when certain metals are illuminated with light of sufficiently short wavelength. Einstein's theory of the photoelectric effect provided some of the first evidence for photons.

Photon A quantum or particle of light. The photon has spin 1 and has no rest mass. The photon is its own antiparticle.

Pion (π^+, π^-, π^0) A meson that carries the force that binds protons and neutrons in the nucleus.

Planck energy (Planck length, Planck time) The energy obtained by combining the three fundamental constants of nature: the speed of light, Planck's constant, and the gravitational constant. Particle collisions that occur at this energy would presumably need to be described by a theory that unifies gravity and quantum mechanics. The same three constants can be combined to get the *Planck length*, which is about 10^{-35} meter, and the *Planck time*, which is about 10^{-43} second.

Planck's constant ($\hbar$) The fundamental constant of quantum mechanics. It is also the fundamental unit of spin.

Positron (e^+) The antiparticle of the electron. It has the same mass and spin as the electron but the opposite electric charge.

Proton A spin 1/2 particle with charge +1 that is one of the three main constituents of ordinary matter. It consists of two up quarks and one down quark.

QED (quantum electrodynamics) The relativistic quantum field theory of electrons, positrons, and photons.

Quantum field The mathematical function that describes the motion of a particle or particles in quantum mechanics and relativistic quantum field theory. The square of the field value gives the probability of detecting a particle at a particular location. In quantum mechanics the quantum field is also known as the *wavefunction*.

Quantum mechanics The theory developed to explain the bizarre behavior of very small objects. Quantum mechanics introduced a new fundamental constant, Planck's constant, and relinquished the certainty of classical mechanics for probability.

Quark An elementary spin 1/2 particle of the Standard Model. Quarks come in six flavors: up, down, charm, strange, top, and bottom. Each flavor comes in three colors: red, blue, and green. Only up quarks and down quarks are present in everyday objects, in the protons and neutrons of atomic nuclei.

Relativistic quantum field theory The extension of the early version of quantum mechanics to make it compatible with special relativity. Except for string theories, all modern elementary particle theories are relativistic quantum field theories.

Relativity / relativistic In this book, these terms always refer to special relativity, Einstein's theory of the behavior of objects moving at high velocity. The basic assumption of the theory is that the speed of light is a constant of nature, independent of the motion of the light source.

Renormalization The procedure for eliminating the infinities that arise in calculations in relativistic quantum field theories. Not all such theories can be renormalized; renormalizability is usually considered a requirement for a viable theory.

Rest mass / rest energy According to special relativity, an object with mass has energy even when it is at rest, in the amount $E = mc^2$. This energy can be released by particle-antiparticle annihilation, for instance. A photon always moves at the speed of light; it has no rest mass.

Scaling A property observed in some high energy particle collisions that gave experimental support to the quark model.

Schrödinger equation The basic equation of quantum mechanics. The solution to this equation is called the quantum field.

Special relativity See Relativity.

Spin Along with mass and electric charge, a fundamental property of elementary particles. It helps to think of a particle as a small ball that spins constantly without ever slowing down. Only the direction of the spin axis can be changed. However, this picture is necessarily incorrect. The spin of a particle cannot be explained in terms of ordinary rotation, it must simply be accepted as a defining property of the particle.

Spontaneous symmetry breaking A sudden change from a symmetrical situation to an asymmetrical one, as when a pencil balanced on its point spontaneously falls.

Strange particles Particles containing at least one strange or antistrange quark.

String theory A class of speculative theories in which elementary particles are not pointlike objects but rather extremely small one-dimensional objects called strings.

Strong force The force that binds protons and neutrons in the atomic nucleus and is responsible for certain types of particle decays. Nowadays used also for the SU(3) color force that binds quarks together.

SU(3) symmetry (SU(3) color force) The symmetry that relates red, blue, and green quarks. The existence of the symmetry explains the color force between quarks, because any theory with such symmetry will necessarily have gluons that mediate the force. The SU(3) symmetry of the Eightfold Way, however, is a different type of symmetry, unrelated to the quark colors.

Superposition state When a particle has a measurable probability of being in two (or more) different quantum states.

Supersymmetry A symmetry that requires a fermion corresponding to every boson, and vice versa. If supersymmetry is unbroken, the fermion and its corresponding boson must have the same mass. If supersymmetry truly describes our universe, it must be a broken symmetry.

Symmetry An operation that leaves an object or theory unchanged, for example, rotating a sphere.

Tau (τ^+, τ^-) The tau, like the muon, is a particle with the same charge and spin as the electron, but with a much larger mass.

Totalitarian theorem "Anything that is not forbidden is compulsory." An expression of the fact that, in relativistic quantum field theory, any particle interaction that is not ruled out by a symmetry or conservation law will happen at least *some* of the time.

Vacuum The state with no particles—in other words, empty space.

Virtual particle An undetectable bit of a relativistic quantum field. Virtual particles can exhibit bizarre behavior such as traveling faster than light, but since they cannot be detected directly, they do not violate special relativity. Though the particles are undetectable, their effects are important; they are the source of the forces between real particles.

W particle (W^+, W^-) Particles predicted to exist by the Standard Model's spontaneous symmetry breaking, and detected for the first time in particle collisions at CERN in 1983. The W particles (and the Z^0) mediate the weak force.

Weak force (weak nuclear force) The force responsible for beta-decay and certain other forms of radioactivity. In the Standard Model, it is simply one aspect of the unified electroweak interaction.

WIMP (weakly interacting massive particle) The generic term for particles that may make up some, or most, of the dark matter in the universe. Although (by definition) they interact weakly with each other and with normal matter, they need not interact via the weak force.

Yang-Mills theory A theory with a continuous symmetry (called a gauge symmetry) such as the SU(3) symmetry of the color force. Also known as a *gauge theory*.

Zeeman effect The splitting of the lines of the hydrogen atom's spectrum when the atom is subjected to a magnetic field.

Z^0 particle Along with the W particle, an intermediate particle of the unified electroweak theory. The Z^0 was detected for the first time in particle collisions at CERN in 1983.

References

Introduction

[1] *Reviews of Modern Physics*, 52 no. 3, p. 1319, cited in R. Crease and C. Mann, *The Second Creation*, p. 253.

[2] *Essays* (1625), quoted in *The Oxford Dictionary of Phrase, Saying, and Quotation*, p. 34.

[3] Quoted in R. Crease and C. Mann, *The Second Creation*, Second Edition, p. 9.

Chapter 1

[1] Quoted in A. Pais, *Subtle Is the Lord*, p. 85.

[2] I. B. Cohen, *Isaac Newton's Papers and Letters on Natural Philosophy*, pp. 302-303.

Chapter 2

[1] A. Pais, *Subtle Is the Lord*, p. 152.

[2] A. Pais, *Subtle Is the Lord*, p. 149., this author's translation

[3] A. Dick, *Emmy Noether: 1882-1935*, p. 30.

[4] J. K. Brewer and M. K. Smith, *Emmy Noether: A Tribute to Her Life and Work*, p. 35.

Chapter 3

[1] R. Feynman, *QED: The Strange Theory of Light and Matter*, p. 9.
[2] Leon Lederman, *The God Particle*, p. 167.

Chapter 5

[1] Nobel Lecture, quoted in J. Gleick, *Genius*, p. 122.

Chapter 6

[1] R. Crease and C. Mann, *The Second Creation*, p. 130.
[2] S. Schweber, *QED and the Men Who Made It*, p.292.
[3] S. Schweber, *QED and the Men Who Made It*, p.334.
[4] S. Schweber, *QED and the Men Who Made It*, p. 303
[5] R. Crease and C. Mann, *The Second Creation*, pp. 70-71.
[6] R. Crease and C. Mann, *The Second Creation*, p. 102.
[7] R. Feynman, *QED: The Strange Theory of Light and Matter*, p. 10.

Chapter 7

[1] Quoted in Leon Lederman, *The God Particle*, p. 153.

Chapter 8

[1] G. Johnson, *Strange Beauty*, p. 119.

[2] G. Johnson, *Strange Beauty*, p. 168.

[3] Caltech report CTSL-20, quoted in R. Crease and C. Mann, *The Second Creation*, p. 267.

[4] G. Johnson, *Strange Beauty*, p. 121.

[5] Y. Ne'eman and Y. Kirsh, *The Particle Hunters*, p. 203.

[6] G. Johnson, *Strange Beauty*, p. 215.

[7] G. Johnson, *Strange Beauty*, p. 236.

[8] G. Johnson, *Strange Beauty*, p. 227.

[9] R. Crease and C. Mann, *The Second Creation*, p. 307.

[10] L. Lederman, *The God Particle*, p. 9.

Chapter 9

[1] A. Pickering, *Constructing Quarks*, p. 178.

Chapter 10

[1] Quoted by E. Salaman, *The Listener*, 54 (1955), pp. 370-371.

[2] R. Crease and C. Mann, *The Second Creation*, p. 248.

[3] R. Crease and C. Mann, *The Second Creation*, p. 209.

[4] L. Lederman, *The God Particle*, p. 390.

Chapter 11

[1] P. Ramond, *Journeys Beyond the Standard Model*, p. 41
[2] Chang-Hwan Lee, Invited Talk at KIAS-APCTP Symposium, Seoul, Korea, November, 2004.

Chapter 12

[1] R. Crease and C. Mann, *The Second Creation*, p. 400.
[2] CERN talk, October 11, 2000.

Further Reading

Many of the topics only touched on in this book are worthy of their own book. Here are some suggestions for where the interested reader can learn more about particular subjects. The following books are all written at a general, non-specialist level.

Crease, Robert P. and Charles C. Mann. *The Second Creation: Makers of the Revolution in Twentieth-Century Physics*. Revised ed. New Brunswick, NJ: Rutgers University Press, 1996. A beautifully written and very thorough history of the Standard Model. If you're only going to read one more book about the Standard Model, make it this one.

Ferris, Timothy. *Coming of Age in the Milky Way*. New York: Anchor, 1988. An excellent account of how we came to our current understanding of stars, galaxies, and the cosmos, from Aristotle to the Big Bang theory.

Feynman, Richard P. *QED: The Strange Theory of Light and Matter*. Princeton, NJ: Princeton University Press, 1985. Feynman's description, in his inimitable style, of the ideas of relativistic quantum field theory as applied to electrons and photons.

Feynman, Richard P., as told to Ralph Leighton. *"Surely You're Joking, Mr. Feynman": Adventures of a Curious Character*, ed. Edward Hutchings. New York: Norton, 1985. Feynman's irreverent stories about his life and his work.

Gamow, George. *Mr. Tompkins in Paperback.* New York: Cambridge University Press, 1993. Combining the famous physicist's *Mr. Tompkins in Wonderland* and *Mr. Tompkins Explores the Atom,* this book tells about special relativity and quantum mechanics through the dream-adventures of the hero.

Gleick, James. *Genius: The Life and Science of Richard Feynman.* New York: Pantheon, 1992. A good biography of the man, including the less funny parts of his life that Feynman left out of *Surely You're Joking.*

Greene, Brian. *The Elegant Universe: Superstrings, Hidden Dimensions, and the Quest for the Ultimate Theory.* New York: Norton, 1999. Greene is amazingly successful at introducing the general reader to the extremely mathematical and complex theory of superstrings.

Guth, Alan H. *The Inflationary Universe: The Quest for a New Theory of Cosmic Origins.* Reading, MA: Addison-Wesley, 1997. A very readable explanation of the big bang theory and cosmic inflation, by the inflation's inventor.

Herbert, Nick. *Quantum Reality: Beyond the New Physics.* New York: Anchor Books, 1985. An introduction to quantum mechanics and the difficulties in its interpretation. Herbert includes a good discussion of the varied philosophical approaches physicists use to deal with the theory, and a very thorough treatment of Bell's theorem and its consequences.

Johnson, George. *Strange Beauty: Murray Gell-Mann and the Revolution in Twentieth-Century Physics.* New York: Knopf, 1999. Another terrific biography, of one of the leading figures in the story of elementary particle physics.

Lederman, Leon, with Dick Teresi. *The God Particle: If the Universe Is the Answer, What Is the Question?* New York: Delta, 1993. The funniest book ever written on elementary particle physics, by one of the Standard Model's leading experimentalists and 1988 Nobel Prize winner.

Ne'eman, Yuval, and Yoram Kirsh. *The Particle Hunters*. Second ed. New York: Cambridge University Press, 1996. Read this for many more details on the particles of the Standard Model and the experiments that discovered them.

Schwartz, Joseph and Michael McGuinness. *Einstein for Beginners*. New York: Pantheon, 1979. This little book lays out the basic ideas of special relativity. Written in comic book format for the general reader, this book nevertheless introduces some of the mathematics of the theory in a very understandable way.

Acknowledgments

The author would like to thank Rabi Mohapatra and Tom Cohen of the University of Maryland for illuminating discussions. Many thanks also to Anna Rain, Jeremy Warner, and Michael Slott for their invaluable remarks on an earlier draft of the book.

Index